Lecture Notes in Mathematics

Edited by A. Dold and B. Eckmann

781

Harmonic Analysis Iraklion 1978

Proceedings of a Conference Held at the University of Crete, Iraklion, Greece, July 1978

Edited by
N. Petridis, S. K. Pichorides, and N. Varopoulos

Springer-Verlag
Berlin Heidelberg New York 1980

Editors

Nicholas Petridis
Eastern Illinois University
Department of Mathematics
Charleston, IL 61920
USA

Stylianos K. Pichorides
N. R. C. Demokritos
Aghia Paraskevi
Attikis
Greece

Nicolas Varopoulos
Department of Mathematics
University of Paris XI
Orsay 91
France

AMS Subject Classifications (1980): 42-XX, 43-XX

ISBN 3-540-09756-2 Springer-Verlag Berlin Heidelberg New York
ISBN 0-387-09756-2 Springer-Verlag New York Heidelberg Berlin

Library of Congress Cataloging in Publication Data. Symposium on Harmonic Analysis, University of Crete, 1978. Harmonic analysis, Iraklion 1978. (Lecture notes in mathematics; 781) Bibliography: p. Includes index. 1. Harmonic analysis--Congresses. I. Petridis, N. II. Pichorides, S. K., 1940- III. Varopoulos, N., 1940- IV. University of Crete. V. Title. VI. Series: Lecture notes in mathematics (Berlin); 781. QA3.L28 no. 781 [QA403] 510s [515'.2433] 80-10989
ISBN 0-387-09756-2

Printed in Germany

Printing and binding: Beltz Offsetdruck, Hemsbach/Bergstr.
2141/3140-543210

FOREWORD

This volume represents the talks delivered at the Symposium on Harmonic Analysis, held at the University of Crete, in Iraklion, Greece, the first week of July 1978.

The conference was organized by the newly founded University of Crete on the occasion of its first anniversary.

The manuscripts of the lectures are published here, as supplied to us by the speakers, except for retyping to make them uniform in appearance.

The common feature of these lectures is that either they strictly belong to Harmonic Analysis (classical and abstract) or they use methods belonging to it.

We believe that we express the feelings of all participants if we extend our thanks not only to our host, the University of Crete, but also to a number of local communities (Iraklion, Aghios Nikolaos, Acharnes, Anogia, etc.) which transformed their love for and their faith in the new University to an unforgettable hospitality for its guests.

We also wish to thank

- The Ministry of Science and Culture,
- The Mayor and the Town Council of Iraklion,
- The National Tourist Organization of Greece,

for financial support.

The Editorial Committee

N. Petridis, S. Pichorides, N. Varopoulos

CONTENTS

+ To appear elsewhere.

CRITERIA FOR ABSOLUTE CONVERGENCE OF FOURIER SERIES

Nicolas K. Artémiades

To give, even a partly expository, talk to an audience containing such a number of experts does not seem to be an easy task. I am afraid certain people will hear the speaker explaining their theorems to them. But that will just have to be.

1. INTRODUCTION

One of the primary objectives in the theory of Fourier series is the study of the class A of all Lebesgue integrable complex-valued functions on the circle T (the additive group of the reals modulo 2π) whose Fourier series are absolutely convergent. Let A be the set of all continuous functions which belong to A. It is well known that A is a Banach space under pointwise linear operations and the norm $||f||_A = \sum_{n\in Z} |\hat{f}(n) < +\infty$, where $\hat{f}(n)$ is the nth Fourier coefficient of the function $f \in A$. Also, A is an algebra under pointwise multiplication and $||fg||_A \leq ||f||_A\,||g||_A$. This means that A is a commutative Banach algebra with the constant function 1 as its identity element.

The Banach Algebra structure of A (due to N. Wiener) suggests a great number of problems which constitute the so called "modern approach" to the study of A. For example, one of these problems on which attention has been concentrated is to find "under what conditions upon the function F, defined on some subset D of the complex plane, is it true that $F \circ f \in A$ whenever $f \in A$ and $f(\mathbb{R})$ D ?" Another problem is to"determine the closed ideals of A". A particular case of this last problem is the so called "Problem of Spectral Synthesis" which can be formulated as follows: "Is every closed ideal of A of the form I_E, where I_E is the closed ideal of A formed by all functions in A which vanish on the closed set E ?" A negative answer to this question was given in 1959 by P. Malliavin.

But I will not continue further towards the direction suggested by the Banach Algebra structure of A.

The classical approach to the study of A has in the main concentrated attention on seeking conditions on an individual function f, which are sufficient and/or necessary to ensure that $f \in A$. Wiener proved that the property of a continuous function on T to belong to A is a local one. This simply means that if f is continuous on T and if for each $a \in T$ there exists a function $g_a \in A$ which is equal to f in some neighborhood of a, then $f \in A$. In the classical approach, emphasis is given to comparing this local property to other properties, as for example is the modulus of continuity. Into this direction of research fall developments concerning the restrictions of the class A, noted A(E), to closed subsets E of the circle. There are closed subsets of the circle (called Helson sets) such that every continuous function on E belongs to A(E). In general, it is true that more E is "fat" more severe is the condition to be imposed on the modulus of continuity of a function f in order that $f \in A(E)$.

In many instances, the study of a problem in A is facilitated if it is transferred to an A(E) for a certain E.

To finish with this very brief expository part of the article I would like to mention two well known criteria for a function f to be in .

<u>Criterion of M. Riesz</u>. $f \in A$ iff it can be expressed in the form $f = u*\upsilon$ with $u,\upsilon \in L^2(T)$.

Unfortunately, this criterion is very difficult to apply in any specific case that is not already decidable on more evident grounds.

<u>Steckin's Criterion</u>. For every $f \in L^2(T)$ and every integer $n \geq 0$ set

$$e_n(f) = \inf ||f-P||_{L^2(T)}$$

where the infimum is taken over all trigonometric polynomials with at most n coefficients different from zero ($P(t) = \sum_{m=1}^{n} \gamma_m e^{i\lambda_m t}$). We have $\sum_{n \in Z} |\hat{f}(n)| < +\infty$ iff $\sum_{n=1}^{\infty} \frac{1}{\sqrt{n}} e_n(f) < +\infty$.

The main drawback with this criterion is the extreme difficulty encountered in estimating the numbers $e_n(f)$ for a given function.

2. Some other criteria

It is well known (Kahane [2], p. 9) that "every continuous function on T with non negative Fourier coefficients belongs to A".

This result can easily be generalized as follows:

Theorem 1. Let $f: T \to \mathbb{C}$ be continuous with the property that there is a $a \in \mathbb{R}$ such that $a \leq \arg \hat{f}(n) \leq a + \frac{\pi}{2}$ ($n \in Z$). Then $f \in A$.

Also every $f \in A$ is a linear combination of continuous functions on T with the above property.

Proof. The second part of the theorem is obvious. To prove the first part let us assume, without loss of generality, that $a = 0$. For if $a \neq 0$ we may consider $g(x) = f(x) \cdot e^{-ia}$ instead of f, since $0 \leq \arg \hat{g}(n) \leq \frac{\pi}{2}$. Next, set

$$F(x) = \frac{f(x) + \overline{f(-x)}}{2}, \quad G(x) = \frac{f(x) - \overline{f(-x)}}{2i}.$$

Clearly, both F and G are continuous and $\hat{F}(n) = \mathrm{Re}\hat{f}(n) \geq 0$, $\hat{G}(n) = \mathcal{J}m\hat{f}(n) \geq 0$, so that by the previous result F and G belong to A which means $f \in A$ since A is a linear space.

Theorems 2 and 3 below provide criteria for f to be in A.

Theorem 2. Let $f \in L^1(T)$. Then $f \in A$ iff the following condition is satisfied.

"There is a Lebesgue point, a, for f such that the sequences

(*) $\langle (\operatorname{Re} \hat{f}_a(n))^- \rangle_{n \varepsilon Z}$, $\langle (\mathcal{J}m \hat{f}_a(n))^- \rangle_{n \varepsilon Z}$

belong to ℓ'.

<u>Proof</u>. Suppose (*) is satisfied. We first consider the case $a = 0$. For N a positive integer set $\sigma_N(t) = \sum_{n=-N}^{N} (1 - \frac{n}{N}) \hat{f}(n) e^{int}$.

By a theorem of Lebesgue we have that $\lim_{N\to\infty} \sigma_N(t) = f(t)$, if f is a Lebesgue point for f. Hence

$$(1) \qquad \lim_{N\to\infty} \sigma_N(0) = \lim \sum_{n=-N}^{N} (1 - \frac{|n|}{N}) \hat{f}(n) = \hat{f}(0) = \text{finite.}$$

Also

$$(2) \qquad {}_N(0) = \sum_{n=-N}^{N} (1 - \frac{|n|}{N}) \hat{f}(n) =$$

$$= \sum_{n=-N}^{N} (1 - \frac{|n|}{N}) (\operatorname{Re}\hat{f}(n))^+ + i \sum_{n=-N}^{N} (1 - \frac{n}{N}) (\mathcal{J}m\hat{f}(n))^+$$

$$- \sum_{n=-N}^{N} (1 - \frac{|n|}{N}) (\operatorname{Re}\hat{f}(n))^- - i \sum_{n=-N}^{N} (1 - \frac{|n|}{N}) (\mathcal{J}m\hat{f}(n))^-.$$

If we let $N \to \infty$ then the $\sigma_N(0)$ are uniformly bounded because of (1), while the last two sums of the right-hand side of (2) are bounded (more precisely they converge because of the hypothesis (*)). Therefore

$$\lim_N \sum_{n=-N}^{N} (1 - \frac{n}{N}) (\operatorname{Re}\hat{f}(n))^+ < +\infty$$

$$\lim_N \sum_{n=-N}^{N} (1 - \frac{n}{N}) (\mathcal{J}mf(n))^+ < +\infty$$

Since the Cesāro summability of a series with non-negative terms implies the convergence of the series, it follows that $\sum_{n\varepsilon Z} |\hat{f}(n)| < +\infty$, i.e. $f \varepsilon A$.

Next assume $a \neq 0$. Then 0 is a Lebesgue point for f_a so that, by the last result, we have $\sum_{n\varepsilon Z} |\hat{f}_a(n)| < +\infty$. But $\hat{f}_a(n) = e^{ian}\hat{f}(n)$ so that $f \in A$.

Now, if $f \in A$ and a is any Lebesgue point for f we have $\sum_{n\in Z} |\hat{f}(n)| = \sum_{n\in Z} |e^{ian}\hat{f}(n)| = \sum_{n\in Z} |\hat{f}_a(n)| < +\infty$ and condition (*) is clearly satisfied.

<u>Corollary 1</u>. Let $f \in L^1(T)$. Then f is equal a.e to a linear combination of positive definite functions iff condition (*) is satisfied.

<u>Proof</u>. It follows from Theorem 2 and Herglotz's characterization of continuous functions with non-negative coefficients as positive definite.

<u>Theorem 3</u>. Let $f \in L^1(T)$. Then $f \in A$ iff the following condition is satisfied:

"f is ess. bounded in a neighborhood of some real number a, and both sequences

(**) $$<(\mathrm{Re}\, \hat{f}_a(n))^- >_{n\in Z}, \quad (\mathcal{J}m\, \hat{f}_a(n))^- >_{n\in Z} \text{ belong to } \ell^1."$$

<u>Proof</u>. Suppose (**) is satisfied and let $a = 0$. Using the notation of Theorem 2 we have:

$$\sigma_N(t) = \frac{1}{2} \int_T f(y) K_N(t-y)\, dy$$

where $K_N(y) = \sum_{n=-N}^{N} (1 - \frac{|n|}{N}) e^{iny} = \frac{\sin^2(N|2)y}{N \sin^2(y|2)}$.

Next assume $|f(y)| \leq M$ a.e for $y \in (-h,h)$ $(h > 0)$.

We have

$$\sigma_N(0) = \frac{1}{2\pi} \int_{-h}^{h} f(y) K_n(y)\, dy + \frac{1}{2\pi} \int_{-h}^{h} (\ ) + \frac{1}{2\pi} \int (\ ).$$

Observe that the first of the last three integrals is bounded by M, and the other two converge to zero as $N \to \infty$ by the Lebesgue dominated convergence theorem. Therefore the $\sigma_N(0)$ are uniformly bounded.

From this point on the proof proceeds exactly the same way as in Theorem 2., that is by letting $N \to \infty$ in (2).

Corollary 2. Let $f \in L^1(T)$. Then f is equal a.e. to a linear combination of positive definite functions iff (**) is satisfied.

Using a technique similar to the one used above one proves the following analogues of theorems 2 and 3.

Theorem 2'. Let $f \in L^1(\mathbb{R})$. Then $\hat{f} \in L^1(\mathbb{R})$ (where $\hat{f}$ is the Fourier transform of f) iff there is a Lebesgue point a for f such that $(\mathrm{Re}\hat{f}_a)^-$ $(\mathcal{J}m\,\mathrm{Re}\hat{f}_a)^-$ belong to $L^1(\mathbb{R})$.

Theorem 3'. Let $f \in L^1(\mathbb{R})$. Then $\hat{f} \in L^1(\mathbb{R})$ iff f is essentially bounded in a neighborhood of some real number a and $(\mathrm{Re}\hat{f}_a)^-$, $(\mathcal{J}m\hat{f}_a)^-$ belong to $L^1(\mathbb{R})$.

One might find theorems 2 and 3 interesting also because of the following remark.

Remark

Call a numerical series $\Sigma a_n + ib_n$, $(a_n, b_n \in \mathbb{R})$ "one sidedly absolutely convergent" (O.A.C.), iff: (at least one of Σa_n^+, Σa_n^-) and (at least one of Σb_n^+, Σb_n^-) is finite.

Now, it is possible that a series $\Sigma(a_n + ib_n)$ is not OAC while the series $\Sigma(a_n + ib_n)e^{i\lambda}$ is OAC. In other words a non OAC series can, in some cases, be converted to an OAC series by just multiplying each term by a factor of the form $e^{i\lambda}$ (λ = some constant) or perhaps in some other way.

Example: Let $c_n = a_n + ib_n$ where $c_{2n} = 1 + i$, $c_{2n+1} = 1 - i$, $n = 0,1,2,\ldots$ and $\lambda = \frac{\pi}{4}$. Then it is easily seen that Σc_n is not OAC while $\Sigma c_n e^{i\lambda}$ is.

Theorems 2 and 3 essentially say that the Fourier series of f converges absolutely iff $\Sigma\hat{f}_a(n)$ is OAC.

REFERENCES

[1] Artémiades, N. "Criteria for absolute convergence of Fourier series", Proc. Amer. Math. Soc. 50 (1975) 179-183.

[2] Kahane, J.-P., "Séries de Fourier absolument convergentes", Ergebnisse der Math. 50, Springer-Verlag (1970).

[3] Zygmund, A., Trigonometric Series (2nd edition) 2 vols., Cambridge, England, 1959.

FRACTIONAL CARTESIAN PRODUCTS IN HARMONIC ANALYSIS

by

Ron C. Blei(*)

The Hebrew University and The University of Connecticut

Our purpose here is to explain the fractional cartesian products of [1] which naturally filled gaps between ordinary cartesian products of sets in a framework of harmonic analysis. The idea of fractional powers of sets appears fairly general and we would first like to describe briefly--taking a somewhat metamathematical point of view--the philosophy behind these products. Let E be a given set. Let L be a positive integer, Y be a fixed indexing space and $\{f_i\}_{i=1}^{L}$ be a collection of functions from Y onto E. Consider now the following subset of the usual L-fold product of E:

$$E_{(f_i)} = \{(f_i(y))_{i=1}^{L} : y \in Y\} \subset E^L$$

If the f_i's are 'independent' (for any $x_1, \ldots, x_L \in E$, the system of equations $f_i(y) = x_i$, $i = 1, \ldots, L$, has a solution in Y) then $E_{(f_i)} = E^L$. On the other extreme, if the f_i's are mutually 'dependent'

$$(f_i(y_1) = f_i(y_2) \Rightarrow f_j(y_1) = f_j(y_2) \text{ for any } f_i, f_j \text{ and } y_1, y_2 \in Y)$$

then $E_{(f_i)}$ can be canonically identified with E. If, however, the type of interdependencies between the f_i's falls somewhere between independence and mutual dependence, $E_{(f_i)}$ is then a set that falls somewhere between E and E^L.

To see how to formulate an intermediate type of interdependencies we observe that independence and dependence can be measured in

(*) Author was supported partially by NSF Grant MCS 76-07135.

the following way. First, by replacing Y with an appropriate quotient of Y, we can assume without loss of generality that $y \to (f_i(y))_{i=1}^{L}$ is an injection. Let s be a positive integer, and $A_1,\ldots,A_L \subset E$ be arbitrary where $|A_1| = \cdots = |A_L| = s$ ($|\cdot|$ denotes cardinality). Write:

$$\phi_{(A_i)}(s) = |\{y\varepsilon Y : f_1(y)\varepsilon A_1 \text{ and } \ldots \text{ and } f_L(y)\varepsilon A_L\}|.$$

Note that if the f_i's are independent then

$$\phi_{(A_i)}(s) = s^L;$$

on the other hand, if the f_i's are mutually dependent then

$$\phi_{(A_i)}(s) \leq s.$$

An intermediate interdependency for $\{f_i\}_{i=1}^{L}$ that corresponds to $1 < r < L$ can be described by the relation (asymptotic in s).

$$(1) \quad \psi(E_{(f_i)};s) = \sup\{\phi_{(A_i)}(s) : A_1,\ldots,A_L \subset E,\ |A_1|,\ldots,|A_L| \leq s\} \sim s^r.$$

This is the basic idea underlying the fractional products of [1] where prescribed interdependencies between concrete f_i's simulated the desired fractional power of a set.

We now move to a harmonic analytic context, where we start with $E = \{\gamma_i\}_{i=1}^{\infty}$, an infinite independent set in some discrete abelian group Γ; that is, for any L, L' > 0 the relation

$$\prod_{j=1}^{L} \gamma_j^{\lambda_j} = \prod_{j=1}^{L'} \gamma_j^{\nu_j}$$

where the λ_j's and ν_j's are arbitrary integers, implies that $L = L'$ and $\lambda_j = \nu_j$ for all j. For example, E could be the canonical basis in $\bigoplus Z$ (the infinite direct sum of Z) whose compact dual group is $\bigotimes T$ (the infinite direct product of T). We proceed to construct a fractional cartesian product of E. Let $J \geq K > 0$ be given integers, and let

$$\mathcal{J} = \{1,\ldots,J\}.$$

For the sake of typographical convenience, we write $N = \binom{J}{K}$. Let

$$\{S_1,\ldots,S_N\}$$

be the collection of all K-subsets of $\mathcal{J}$ (sets containing K elements of $\mathcal{J}$), where each $S_\alpha \subset \mathcal{J}$ is enumerated as

$$S_\alpha = (\alpha_1,\ldots,\alpha_K).$$

Let $P_1,\ldots,P_N$ be the projections from $(Z^+)^J$ onto $(Z^+)^K$ defined as follows: For $1 \leq \alpha \leq N$ and $j = (j_1,\ldots,j_J) \in (Z^+)^J$,

$$P_\alpha(j) = (j_{\alpha_1},\ldots,j_{\alpha_K}).$$

Next, let f be any one-one function from $(Z^+)^J$ onto E, and

$$f_\alpha = f \circ P_\alpha : (Z^+)^J \to E;$$

write

$$E_{(f_\alpha)} = E_{J,K} = \{(f_1(j),\ldots,f_N(j)) : j \in (Z^+)^J\} \subset E^N \subset \Gamma^N.$$

The outstanding feature of $E_{J,K}$ is that

$$\psi(E_{J,K};s) \sim s^{J/K}$$

(see (1) for the definition of ψ), which is, in fact, an analogue of a basic harmonic analytic (or probabilistic) property of $E_{J,K}$ that will now be discussed. First, we recall that a spectral set $F \subset \Gamma$ is a $\Lambda(p)$ set, $2 < p < \infty$, if there is a constant $A > 0$ so that for all functions $h \in L^2(G)$ whose spectrum is in F $(G = \Gamma\hat{})$, we have

$$(2) \qquad A \| h \|_2 \geq \| h \|_p .$$

The 'smallest' A for which (2) holds is the $\Lambda(p)$ constant of F and is denoted by $A(p,F)$.

<u>Definition</u>. Let $\beta \in [1,\infty)$ $F \subset \Gamma$ is a Λ^β set if $A(p,F)$ is $\mathcal{O}(p^{\beta/2})$. F is said to be <u>exactly-Λ^β</u> when F is Λ^a if and only if $a \in [\beta,\infty)$, and <u>exactly non-Λ^β</u> when F is Λ^a if and only if $a \in (\beta,\infty)$.

J-fold cartesian products of independent sets are the prototypical examples that are exactly Λ^J (see [2]). The gaps that were kept open between J and J+1 are neatly filled, as we are about to see, by the fractional products that have just been defined.

<u>Theorem</u>. Let $E \subset \Gamma$ be an independent set. Then, $E_{J,K} \subset \Gamma^N$ is exactly $\Lambda^{J/K}$, and, moreover, there is $\eta_{J,K} > 0$ so that for all $q > 2$

$$(*) \qquad \eta_{J,K}\, q^{J/2K} \leq A(q, E_{J,K}) \leq q^{J/2K} .$$

To avoid a fog of indices, we sketch the proof of the theorem in the case $J = 3$ and $K = 2$ -- the general case follows a similar line.

The right hand inequality in (*) is based on a simple combinational criterion that is a link between the algebraic structure of a spectral set and its harmonic analytic features. Let F be a subset of Γ, s be a positive integer and $\gamma \in \Gamma$. Let $r_s(F,\gamma)$ denote the number of ways to write γ in the form of

(3) $$\gamma = \gamma_1 \cdots \gamma_s ,$$

where $\gamma_1, \ldots, \gamma_s$ are (not necessarily distinct) elements in F, and where different permutations on the right hand side of (3) are counted as different representations. An application of Plancherel formula and the Schwartz inequality yields

(4) $$A(2s,F) \leq \sup \{ [r_s(F,\gamma)]^{1/2s} : \gamma \in \Gamma \}$$

(see Théorème 3 in [2]).

We now present $E_{3,2} \subset \Gamma^3$ as

$$E_{3,2} = \{ [\gamma_{ij}, \gamma_{jk}, \gamma_{ik}) : i,j,k = 1, \ldots \}$$

where $\{\gamma_{ij}\}_{i,j=1}^{\infty}$ is some fixed enumeration of our independent set E, and proceed to estimate $r_s(E_{3,2},\delta)$ for any given $(\delta_1,\delta_2,\delta_3) = \delta \in \Gamma^3$. Suppose that

(5) $$(\delta_1,\delta_2,\delta_3) = (\prod_{n=1}^{s} \gamma_{i_n j_n} , \prod_{n=1}^{s} \gamma_{j_n k_n} , \prod_{n=1}^{s} \gamma_{i_n k_n}) .$$

The independence of E implies that the only way that δ can be obtained as a product of s elements from $E_{3,2}$ is for these elements to have in their first, second and third coordinates the members of E that appear in the first, second and third coordinates of (5), respectively. Let

$$A_1 = \{ (i_1 j_1), \ldots, (i_s j_s) \} ,$$

$$A_2 = \{ (j_1 k_1), \ldots, (j_s k_s) \} ,$$

$$A_3 = \{ (i_1 k_1), \ldots, (i_s k_s) \} ,$$

and

$$V = \{(i,j,k) : (ij) \in A_1, (jk) \in A_2, \text{ and } (ik) \in A_3\} .$$

By virtue of the preceding remark, we have that

(6) $$r_s(E_{3,2},\delta) \leq |V|^s .$$

But,

(7) $$|V| = \sum_{i,j,k=1}^{\infty} \chi_{A_1}(i,j)\ \chi_{A_2}(j,k)\ \chi_{A_3}(i,k) ,$$

where χ_{A_m} is the characteristic function of A_m, $m = 1,2,3$. Applying the Schwartz inequality to the right hand side of (7), we deduce

(8) $$|V| \leq \|\chi_{A_1}\|_2\ \|\chi_{A_2}\|_2\ \|\chi_{A_3}\|_2 \leq s^{3/2} ,$$

and, combining (4), (6) and (8), we obtain

$$A(2s,E_{3,2}) \leq s^{3s/4} .$$

We proceed to verify the left hand inequality in (*). Let $n > 0$ be arbitrary, and g_n be the trigonometric polynomial defined by

$$g_n = \sum_{i,j,k=1}^{n} (\gamma_{ij},\gamma_{jk},\gamma_{ik}) .$$

Next, let h_n be the Riesz product defined by

$$h_n = [\prod_{i,j=1}^{n} (1+\cos(\gamma_{ij},e,e))][\prod_{i,j=1}^{n} (1+\cos(e,\gamma_{ij},e))][\prod_{i,j=1}^{n} (1+\cos(e,e,\gamma_{ij}))],$$

where $\cos(\gamma) = (\gamma+\bar{\gamma})/2$, and e is the identity element in Γ. Observe that $\|h_n\|_1 = 1$ and $\|h_n\|_2 \le \|h_n\|_\infty \le 8^{n^2}$. Therefore, for any $1 < p < 2$

$$\|h_n\|_p \le 8^{n^2/q}, \qquad (1/p + 1/q = 1).$$

Also, observe that $\hat{h}_n = 1/8$ on $\{(\gamma_{ij},\gamma_{jk},\gamma_{ik})\}_{i,j,k=1}^{n}$, and therefore

$$n^3/8 = g_n * h_n(o) \le \|g_n\|_q \, \|h_n\|_p .$$

Letting $q = n^2$, and noting that $\|g_n\|_2 = n^{3/2}$, we finally deduce

$$\frac{n^{3/2}}{16}\,(\|g_n\|_2) \le \|g_n\|_{n^2}$$

and thus obtain the left hand inequality in (*).

Given any $\beta \in [1,\infty)$, we can construct a set that is exactly Λ^β and one that is exactly non-Λ^β by gluing together sufficiently 'thick' pieces of E_{J_n,K_n} where $(J_n/K_n)_{n=1}^{\infty}$ is a monotone increasing sequence (to produce the exact Λ^β property) or decreasing sequence (to produce the exact non-Λ^β property). An interesting open question in this area is the following. For rational $\beta = J/K \in [1,\infty)$ we obtain a set for which (*) holds, and in fact, through the use of Riesz products, we deduced that

$$\eta_{J,K} \ge C^{-(\frac{J}{K})},$$

for some $C > 0$. For an arbitrary β, while the gluing procedure produces a set that is exactly Λ^β, it fails -- in view of (9) -- to produce a set for which (*) holds.

Problem. Is there C > 0 so that

$$\eta_{J,K} \geq C^{-J/K} \quad ?$$

References

1. R. Blei, Fractional cartesian products of sets, Ann. Inst. Fourier (Grenoble).

2. A Bonami, Etude des coefficients de Fourier des fonctions de $L^p(G)$, Ann. Inst. Fourier (Grenoble) 20 (1970), 335-402.

ON A REGULARITY CONDITION FOR GROUP ALGEBRAS OF NON ABELIAN LOCALLY COMPACT GROUPS

J. Boidol

Let G be a locally compact group, $\mathrm{Prim}_* L^1(G)$ the set of the kernels of topological irreducible $*$-representations of $L^1(G)$ on Hilbert space and $\mathrm{Prim}\, C^*(G)$ the set of primitive ideals of the group C^*-algebra $C^*(G)$, both equipped with the Jacobson topology. Then we have a canonical map

$$\Psi : \mathrm{Prim}\, C^*(G) \to \mathrm{Prim}_*\, L^1(G)$$
$$J \to J \cap L^1(G)$$

which is continuous and surjective. One can ask now for those groups, for which this mapping is a homeomorphism. *Let* $[\Psi]$ *denote the class of all these groups.* In [2] the following results were proved:

(A) *Every group* G *which has polynomial growth is in* $[\Psi]$.

(B) *Every group* $G = H \ltimes N$ *with separable abelian groups* H *and* N, *such that the orbit space* $\hat{N}/H$ *is* T_0 *is in* $[\Psi]$.

(C) *Every group in* $[\Psi]$ *is amenable.*

Furthermore it was proved that all connected simply connected solvable Lie groups $G = \exp g$ with $\dim g \leq 4$, $g \neq g_{4,9}(0)$ in the notation of [1, p. 180/181] are in $[\Psi]$

We prove now that $G_{4,9}(0) = \exp g_{4,9}(0)$ *is not in* $[\Psi]$. For doing this we use the method of generalized L^1-algebras developed by Leptin in [4]. Using arguments like Leptin and Poguntke in [5, th. 5] where a certain type of generalized L^1-algebras is shown to be nonsymmetric, we show that algebras of this type are not in $[\Psi]$. (Here we say that a Banach-$*$-Algebra A is in $[\Psi]$, if $\Psi : \mathrm{Prim}_* A \to \mathrm{Prim}\, C^*(A)$ is a homeomorphism.) The algebra $L^1(G_{4,9}(0))$ having such an algebra as quotient is then also not in $[\Psi]$.

1. Let G be a locally compact group, $\Delta : G \to \mathbb{R}^*$ a continuous homoemorphism and $\omega : G \to \mathbb{R}^*$ a symmetric weight on G with $\omega(x) \geq \Delta(x)^{1/2}$ for all $x \in G$. The kernel H of Δ is then a closed normal subgroup of G and $\dot{\omega}(\dot{x}) = \inf_{\xi \in H} \omega(x\xi)$ is a symmetric weight on $G^\cdot = G/H$. According to [7, ch. 3, §7,4] $L^1(G^\cdot,\omega^\cdot)$ is an involutive and commutative Banach algebra and a quotient of $L^1(G,\omega)$. Now $\Delta^\cdot(x^\cdot) = \Delta(x)$ is a well defined continuous homeomorphism from $G^\cdot$ to $\mathbb{R}^*$ and

$$\omega^\cdot(x^\cdot) = \inf_{\xi \in H} \omega(x\xi) \geq \inf_{\xi \in H} \Delta(x\xi)^{1/2} = \Delta^\cdot(\dot{x})^{1/2} .$$

Lemma 1: *Let G be a locally compact group with symmetric weight ω, $\Delta : G \to \mathbb{R}^*$ a continuous homeomorphism such that $\omega(x) \geq \Delta^{1/2}(x)$ for all $x \in G$. If Δ is not trivial, then $L^1(G,\omega)$ is not in $[\Psi]$.*

Proof: Since Banach-*-algebras are not in $[\Psi]$ if they have quotients which are not in $[\Psi]$, it is sufficient by the remarks above to prove the lemma for abelian G. In order to do this we use the criterion of Beurling-Domar (see [3]). We have

$$\sum_{n=1}^{K} \frac{\log \omega(x^n)}{n^2} \geq \sum_{n=1}^{K} \frac{\frac{n}{2} \log \Delta(x)}{n^2} = \frac{1}{2} \log \Delta(x) \sum_{n=1}^{K} \frac{1}{n} .$$

It follows that for $\Delta(x) > 1$ the series $\sum_{n=1}^{\infty} \frac{\log\omega(x^n)}{n^2}$ diverges and that $L^1(G,\omega)$ is not regular.

2. Let now A be a Banach *-algebra with bounded approximating identity. Let A^b denote the adjoint algebra of A. For every $\pi \in \hat{A}$ there exists a canonical extension to a representation π^b of A^b and $\pi \to \pi^b$ induces a homeomorphism from $\mathrm{Prim}_* A$ onto $\mathrm{Prim}_* A^b \setminus h(A)$ where $h(A) = \{J \in \mathrm{Prim}_* A^b \mid A \subseteq J\}$. If $p \in A^b$ is a hermitian idempotent, that

is $p^2 = p = p^*$, then $B = pAp$ is a closed involutive subalgebra of A.

Lemma 2: *If* $p \in A^b$ *is a hermitian idempotent with* $\overline{ApA} = A$, *then*

$$J \to J_p = J \cap pAp, \quad J \in \mathrm{Prim}_* A$$

defines a continuous injective mapping

$$\tau : \mathrm{Prim}_* A \quad \mathrm{Prim}_*\, pAp$$

and τ *is a homeomorphism onto its image.*

Proof: Let $J \in \mathrm{Prim}_* A$, $J = \mathrm{Kern}\ \pi$, $\pi \in \hat{A}$, $X = H_\pi \cdot \pi^b(p)$ is not trivial and because of the irreducibility of $\pi, \pi|_{pAp}$ is not trivial. $\pi|_{pAp}$ gives an irreducible representation on $\pi^b(p)X$ with kernel J_p. Therefore τ is well defined. We have for every twosided ideal $J \leq A$ $J_p = pJp$ and $J = \overline{AJA} = \overline{\overline{ApA}\ J\ \overline{ApA}} = \overline{Ap\ Jp\ A} = \overline{A\ J_p\ A}$. It follows that τ is a homeomorphism onto its image.

Lemma 3: *Let* Q *be a set of hermitian idempotents in* A^b *satisfying the conditions of the Lemma in* [6]. *If for a* $p \in Q$ $\overline{ApA} = A$ *then* $\tau : \mathrm{Prim}_* A \to \mathrm{Prim}_* pAp$ *is a homeomorphism.*

Proof: We remark first that under the conditions of lemma 3 every continuous positive functional f_o on pAp has an extension to a continuous positive functional f on A which is defined by $f(x) = f_o(pxp)$ (see [6]). Let f_o be pure. Then $g \leq f$ implies $\lambda f(pxp) = g(pxp)$ and $g(x^*x) \leq {}^1/\lambda\ g(px^*xp)$ for a $\lambda \in (0,1)$. Let now $q_1,\ldots,q_n \in Q$, $q_1 = p$, $q_i q_j = \delta_{ij}\ q_i$. Then $g(q_i x q_j) = 0$ if $i \neq 1$ or $j \neq 1$. But then $g(x) = g(pxp)$ if $x \in \sum_{i,j=1}^{n} q_i A q_j$ and because of the density of $\sum_{q,q' \in Q} q A q'$ in A $g(x) = g(pxp)$ for all $x \in A$. But this means that $g(x) = \lambda f(x)$ for all $x \in A$. It follows that f is pure. Let now $J_o \in \mathrm{Prim}_* pAp$. Since $p\overline{AJ_oA}p = J_o$ we have to show that $\overline{AJ_oA} \in \mathrm{Prim}_* A$. Let $J_o = \mathrm{kern}\ \pi_o$ $\pi_o \in \hat{A}$. Let f_o be a pure positive functional associated to π_o and f the pure positive functional with $f(x) = f_o(pxp)$ which extends f_o. Let π be the irreducible representation of A associated to f_o, $J = \text{kern}\,\pi$. Then we have

$$\begin{aligned} J \cap pAp &= \{ x \varepsilon pAp \mid f(Ax^*xA) = 0 \} \\ &= \{ x \varepsilon pAp \mid f_o(pApx^*xpAp) = 0 \} \\ &= \text{kern}\,\pi_o = J_o . \end{aligned}$$

It follows from lemma 2 that $J = \overline{AJ_oA}$ and that τ is a homeomorphism. Let now Q be a set of hermitian idempotents in A^b satisfying the conditions of the Lemma in [6]. Let $p \varepsilon Q$ with $\overline{ApA} = A$. Then Q may be considered as a subset of $C^*(A)^b$ satisfying the same conditions and with $\overline{C^*(A)pC^*(A)}^{\|\,\|_{C^*(A)}} = C^*(A)$. Furthermore we have $C^*(pAp) = pC^*(A)p$. Therefore we have by Lemma 3 the following theorem:

<u>Theorem 1</u>: *Let* A *be an involutive Banach algebra with bounded approximating identity, let* $Q \subseteq A^b$ *be a set of hermitian idempotents satisfying all the conditions of the Lemma in* [6]. *If* $p \varepsilon Q$ *is a hermitian idempotent with* $\overline{ApA} = A$, *then* $A \varepsilon [\Psi]$ *if and only if* $B = pAp$ *is in* $[\Psi]$.

<u>Proof</u>: Consider the commuting diagram:

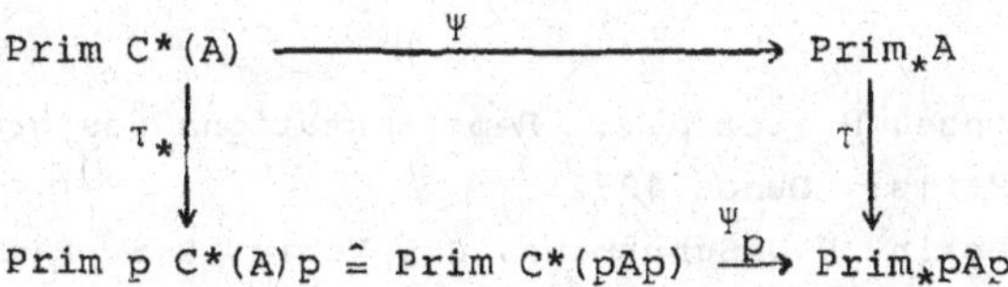

By Lemma 3 τ and τ^* are homeomorphisms. It follows that Ψ is a homeomorphism if and only if Ψ_p is a homeomorphism.

3. Let now G and H be locally compact groups and let $G \times H \to H$, $(g,x) \to x^g \varepsilon H$ be a continuous action of G on H with $(xy)^g = x^g y^g$, $(x^g)^h = x^{gh}$ and $e^h = e$. Let the module Δ of this action (that is $d(x^g) = \Delta(g(dx)$ for a left invariant measure on H) be

not trivial. Let Q be a subalgebra of $C_\infty(H)$ with the properties described in [5, th. 4]. For $g \in G$ and $u \in Q$ let $u \circ g: H \to \mathbb{C}$ be defined by $u \circ g(x) = (x^{g^{-1}})$. We assume that for every $g \in G$ $u \to u \circ g$ is an isometric isomorphism of Q and that $g \to u \circ g$ is for every $u \in Q$ a continuous mapping from G to Q. Then the generalized L^1-algebra $A = L^1(H,Q)$ becomes a G-algebra by $f^g(x) = \Delta(g)^{-1} f(x^{g^{-1}}) \circ g$ and we put $L = L^1(G,A)$.

Theorem 2: $L \notin [\Psi]$

Proof: According to [5, th. 5] there is a system Q of hermitian idempotents in L^b satisfying all the conditions of the Lemma in [6]. Also if $p \in Q$ them $\overline{LpL} = L$ and after Theorem 1 $L \in [\Psi]$ if and only if $pLp \in [\Psi]$. But again after [5, th. 5] $pLp \simeq L^1(G,\omega)$ where ω is a symmetric weight on G with $\omega(x) \geq \Delta(x)^{1/2}$. It follows from Lemma 1 that pLp and therefore L is not in $[\Psi]$.

Corollary: $G_{4,9}(0)$ *is not in* $[\Psi]$.

Proof: After [5] $L^1(G_{4,9}(0))$ has a quotient which is isomorphic to an algebra L as in Theorem 2. Therefore it cannot be in $[\Psi]$.

REFERENCES

[1] Bernat, P., Conze, N., et al.: Représentations des groupes de Lie résolubles. Paris: Dunod 1972.

[2] Boidol, J., Leptin, H., Schürmann, J., Vahle, D.: Räume primitiver Ideale von Gruppenalgebren. Math. Ann. 236, 1-13 (1978).

[3] Domar, Y.: Harmonic analysis based on certain commutative Banach Algebras. Acta Math. 96, 1-66 (1956).

[4] Leptin, H.: Verallgemeinerte L^1-Algebren und projektive Darstellungen lokalkompakter Gruppen. Inventiones math. 3, 257-281, und 4, 68-86 (1967).

[5] Leptin, H., Poguntke, D.: Symmetry and nonsymmetry for locally compact groups. To appear in Journal of Functional Analysis.

[6] Poguntke, D.: Nilpotente Liesche Gruppen haben symmetrische Gruppenalgebren. Math. Ann. 227, 51-59 (1977).

[7] Reifer, H.: Classical harmonic analysis and locally compact groups. Oxford: Clarendon Press 1968.

J. Boidol
Fakultät für Mathematik
der Universität Bielefeld
Universitätsstrasse 1
D-4800 Bielefeld
Federal Rep. of Germany

SINGULAR POSITIVE DEFINITE FUNCTIONS

Alessandro Figà-Talamanca
University of Rome, Rome, Italy
and
University of Maryland, College Park,
Maryland

The purpose of this lecture is to describe a method for constructing positive definite continuous functions on a locally compact unimodular group G. This method was developed in [1] and [2] and it applies to all <u>unimodular</u> locally compact groups the <u>regular representation of which is not completely reducible</u>. In particular it applies to all noncompact Abelian groups. It was used in [1] and [2] to produce examples of positive definite continuous functions which vanish at infinity but are not coefficients of the regular representation.

We will consider in this lecture only the case of an Abelian, locally compact, noncompact group G, with a nondiscrete character group The general case is treated, as in [2], using noncommutative integration theory, with reference to the von Neumann algebra of the regular representation. Although all the essential features of the construction are present in the commutative case, we must remember that certain shortcuts or alternative ways are not allowed if we want to have a proof which extends to the general case. E.g., we cannot freely use the algebraic structure or the topological structure of the character group Γ. We can use, however, the properties of Γ as a measure space with respect to its Haar measure, and the fact that the subalgebra of $L^\infty(\Gamma)$ which consists of the Fourier transforms of elements of $L^1(\Gamma)$, is weak* dense in L^∞.

We shall introduce now some notation. We denote by D the countable product of two element groups $D = \{-1,1\}^\omega$. For $t = \{\varepsilon_j\} \in D$ we define the Rademacher functions $r_j(t) = \varepsilon_j$. In other words $t = \{r_j(t)\}$. We call the products, w_n, of Rademacher functions, Walsh functions and we number them so that $w_n = r_{j_1} r_{j_2} \ldots r_{j_s}$ if $n = 2^{j_1-1} + 2^{j_2-1} + \ldots + 2^{j_s-1}$.

The compact group D is also called the Cantor group.

We shall consider finite positive Borel measures on D, which we call briefly positive measures. The notion of absolute continuity and singularity is always understood with respect to the Haar measure on D. To each positive measure μ on D we will associate a positive definite continuous function φ on G in such a way that if μ is absolutely continuous then φ is the Fourier transform of a positive element of $L^1(\Gamma)$, and conversely if μ is singular, then the positive definite function φ^t, associated to the translated measure μ^t ($\mu^t(A) = \mu(A+t)$), is not the Fourier transform of an element of $L^1(\Gamma)$, for almost all $t \in D$. With a slight modification of the method described here one can obtain that if μ is a product measure on D and $\lim_n \hat{\mu}(w_n) = 0$, then $\varphi \in C_0(G)$. This is what is done in [2].

Let E be a measurable subset of Γ with Haar measure one. A system of Rademacher functions associated with E is any sequence of functions $\{R_j\}$ which are zero off E, are independent as random variables, and take the values 1 and -1 on subsets of E of equal measure. Since the Haar measure on E is nonatomic, we can certainly construct a system of Rademacher functions on E. (Divide E into two subsets of equal measure to define the first Rademacher function, and then devide similarly each of the two subsets to define the second Rademacher function, etc.) We notice that any subsequence R_{j_k} of a Rademacher system R_k is also a Rademacher system relative to E. Starting with a given sequence $\{R_j\}$ one can define a Walsh system $\{W_n\}$ on E, by defining $W_0 = I_E$ to be the indicator function of E and W_n to be a product of Rademacher functions with the same numbering which is defined for the classical Rademacher and Walsh functions on the Cantor group. We now fix a given Rademacher system $\{R_j\}$ which generates a Walsh system $\{W_n\}$, relative to a fixed subset $E \subseteq \Gamma$, of measure one. Each $x \in G$ may be considered as a function (a character) on Γ, $x = x(\gamma)$ and we may thus define

(1) $$g(x) = \sum_n |\hat{W}_n(x)|^2 = \sum_n |\int_\Gamma W_n(\gamma) x(-\gamma)\, d\gamma|^2 .$$

We observe that $g(x)$ is a continuous function. Indeed the function $x \to x(\gamma) I_E(\gamma)$ is continuous from G to $L^2(\Gamma)$ and $g(x)$ is nothing but the square of the L^2-norm of the orthogonal projection of $x(\gamma) I_E(\gamma)$ into the closed subspace of $L^2(\Gamma)$ generated by the functions $\{W_n\}$. This implies that the series in (1) converges uniformly on compact sets (by Dini's theorem). Note that $g(x)$ vanishes off a σ-compact open subgroup of G, this allows us, without loss of generality to assume that G itself is σ-compact. Let K_n be an increasing sequence of compact subsets of G, with the property that every compact set is, for some n, contained in K_n.

We will choose now a sequence of positive integers j_k such that if $R'_k = R_{j_k}$ and $\{W'_n\}$ is the corresponding Walsh system, the series

(2) $$h(x) = \sum_{n=0}^{\infty} |\hat{W}'_n(x)|$$

converges uniformly on compact sets to a continuous (but perhaps unbounded) function. Once the new system R'_k is defined we will carry on our construction using only the Walsh system W'.

Since the series (1) converges uniformly on compact sets, there exists a positive integer i_n such that

(3) $$\sum_{k \geq i_n} |\hat{W}_k(x)|^2 < 2^{-4n} \qquad \text{for } x \in K_n .$$

Let j_n be an increasing sequence such that $i_n < 2^{j_n - 1}$. Define $R'_k = R_{j_k}$. Notice that if m is at least 2^{n-1}, then, by (3),

$$|\hat{W}'_m(x)| < 2^{-2n} \qquad \text{for } x \in K_n .$$

Let K be any compact set and let $K_n \supseteq K$. Then

$$\sum_{k=0}^{\infty} |\hat{W}'(x)| = \sum_{k<2^{n-1}} |\hat{W}_k'(x)| + \sum_{j=n-1}^{\infty} \sum_{k=2^j}^{2^{j+1}-1} |\hat{W}_k'(x)| .$$

If $x \in K$, then $x \in K_j$ for all $j \geq n$. Therefore, $|\hat{W}_k'(x)| < 2^{-2j}$ if $j > n-1$ and $2^j \leq k < 2^{j+1}$. It follows that

$$\sum_{j=n-1}^{\infty} \sum_{k=2^j}^{2^{j+1}-1} |\hat{W}_k'(x)| = \sum_{j=n-1}^{\infty} 2^{-j+1} = 2^{-n+1}$$

This proves that the series (2) converges uniformly on compact sets.

We now define for each $t \in D$

$$p^t(x) = \sum_{k=0}^{\infty} w_k(t)\hat{W}_k'(x),$$

where $w_k(t)$ is the ordinary Walsh function on D. Because of the convergence of (2), the series which define p^t converges uniformly on compact sets and uniformly for $t \in D$. We will show now that $p^t(x)$ is for each t, a positive definite function which therefore is bounded by $1 = p^t(0)$. Indeed the functions

$$p_n^t(x) = \sum_{k<2^n} w_k(t)\hat{W}_k'(x),$$

are the Fourier transforms of the positive functions

$$P_n^t(\gamma) = \prod_{k=1}^{n-1} (I_E + r_k(t)R_k'(\gamma)),$$

which because of the independence of the R_k' have all integral one. Thus for each n and each t, $p_n^t(x)$ is a positive definite function. Since $\lim_n p_n^t(x) = p^t(x)$, we conclude that p^t is also positive definite.

If μ is a positive measure on D we define now a positive definite function φ associated to μ, as

$$\varphi(x) = \int_D p^t(x)\, d\mu(t) = \lim_n \int_D p_n^t(x)\, d\mu(t)$$

$$\sum_{k=0}^{\infty} \hat{\mu}(w_k)\hat{W}_k'(x).$$

It is immediate to verify that φ is positive definite:

$$\sum c_i\bar{c}_j\varphi(x_i-x_j) = \int_D \sum c_i\bar{c}_j p^t(x_i-x_j)\, d\mu(t) \geq 0.$$

If μ is an absolutely continuous measure, then $\varphi(x) = \hat{\Phi}(x)$, with $\Phi \in L^1(\Gamma)$, indeed we may define

$$\Phi_n(\gamma) = \sum_{k<2^n} \hat{\mu}(w_k)W_k'$$

and the convergence in $L^1(D)$ of the sums $\sum_{k<2^n} \hat{\mu}(w_k)w_k(t)$, imply the convergence in $L^1(\Gamma)$ of the functions $\Phi_n(\gamma)$. We would like to establish a partial converse to this fact, but for this more work is needed.

We define a map θ of $L^1(\Gamma)$ onto $L^1(D)$ as follows

$$(\theta F)(t) = \lim_n \sum_{k<2^n} \langle F,W_k'\rangle w_k(t),$$

where $\langle F,W_k\rangle = \int_\Gamma F(\gamma)W_k'(\gamma)\, d\gamma$, and the limit is in the $L^1(D)$ norm. In order to show that the limit exists, we remark first that θ is isometric on the subspace of $L^1(\Gamma)$ which is spanned by $\{W_k'\}$ and that if we define

$$E_n F = \sum_{k<2^n} \langle F,W_k'\rangle W_k,$$

the operators E_n are conditional expectations on $L^1(\Gamma)$ which satisfy $E_nE_m = E_mE_n = E_n$ for $m > n$. Therefore, by the properties of martingales, E_nF converges for each $F \in L^1(\Gamma)$ and so does

$$\theta E_n F = \sum_{k<2^n} \langle F, W_k' \rangle w_k(t),$$

in the norm of $L^1(D)$.

If $\mu \in L^1(D)$ we can find a unique F belonging to the subspace of $L^1(\Gamma)$ generated by the $\{W_k'\}$ such that $\theta F = \mu$. It is sufficient to define $F = \lim_n \sum_{k<2^n} \langle \mu, w_k \rangle W_k'$, where the convergence in $L^1(\Gamma)$ follows because θ is an isometry on the space generated by the $\{W_k'\}$, and partial sums of order 2^n converge in $L^1(D)$. This allows us to define θ also for elements of $L^\infty(\Gamma)$. We let for $G \in L^\infty(\Gamma)$ and $\mu \in L^1$, $\langle \theta G, \mu \rangle = \langle G, F \rangle$, where $\mu = \theta F$ and F is in the closed span of the $\{W_k'\}$. With this definition θ is automatically weak* continuous from $L^\infty(\Gamma)$ to $L^\infty(D)$. If we define

$$\theta_n F(t) = \sum_{k<2^n} \langle F, W_k' \rangle w_k(t),$$

then for $F \in L^1(\Gamma)$, $\lim_n \|\theta_n F - \theta F\|_{L^1(D)} = 0$, and for $F \in L^\infty(\Gamma)$, $\lim_n \theta_n F = F$, weak*. Suppose now that $F \in C_0(\Gamma)$ and $\hat{f} = F$ with f a continuous function with compact support $f \in C_{00}(G)$. Then

$$(\theta_n F)(t) = \int_G p_n^t(x) f(x)\, dx,$$

and since $p_n^t(x)$ converges uniformly for $t \in D$ and x in the support of f, we conclude that $\theta_n F(t)$ converges, uniformly in t, to $\theta F(t)$. It follows, since $\theta_n F \in C(D)$, that $\theta F \in C(D)$. But since $\|\theta F\|_\infty \le \|F\|_\infty$, we also obtain that $\theta F \in C(D)$ for all $F \in C_0(\Gamma)$. We are now ready to state and prove the desired result.

THEOREM. <u>Let</u> μ <u>be a positive singular measure on</u> D. <u>For</u> $t \in D$ <u>define</u> $\mu^t(A) = \mu(A+t)$. <u>Then, for almost every</u> $t \in D$, <u>the positive definite function</u>

$$\varphi^t(x) = \int_D p^{t+s}(x)\, d\,(s) = \int_D p^s(x)\, d\mu^t(s),$$

<u>is not the Fourier transform of an element of</u> $L^1(\Gamma)$.

<u>Proof</u>: Suppose that there exists a set of positive measure $M \subseteq D$, such that for $t \in A$, φ^t is the Fourier transform of an element $\Phi^t \in L^1(\Gamma)$. We will show first that for almost every $t \in M$ and all m,

$$<\Phi^t, W_m> = \hat{\mu}(w_m)w_m(t).$$

For fixed m let $F_\alpha \in C_0(\Gamma)$ be a set of functions converging to W_m in the weak* topology of $L^\infty(\Gamma)$. Then θF_α converges weak* in $L^\infty(D)$ to w_m, and since convolution by a bounded measure is a weak* continuous operator on $L^\infty(D)$

$$\lim_\alpha \theta F_\alpha * \mu = w_m * \mu = \hat{\mu}(w_m)w_m(t).$$

On the other hand for each $t \in M$

$$(\theta F_\alpha * \mu)(t) = \int_\Gamma \Phi^t(\gamma)F_\alpha(\gamma)\, d\gamma,$$

and the last integral converges to $<\Phi^t, W_n>$ for all $t \in M$. This implies that for almost all $t \leftarrow M$

$$<\Phi^t, W_m> = \hat{\mu}(w_m)w_m(t).$$

Since the set $\{W_m\}$ is countable the above equality holds for all m and almost all $t \in M$. Thus, for almost all $t \in M$

$$(\theta_n \Phi^t)(s) = \sum_{k<2^n} \hat{\mu}(w_k)(w_k(t)w_k(s)).$$

But for $t \in M$, $\theta_n \Phi^t$ converges to $\theta\Phi^t$ in $L^1(D)$, which means that the partial sums of order 2^n of the measure μ^t converge in $L^1(D)$. This is impossible since μ^t is singular, and the contradiction proves the theorem.

References

[1] A. Figà-Talamanca. Positive definite functions which vanish at infinity, Pac. J. of Math. 69 (1977), 355-363.

[2] A. Figà-Talamanca. Construction of positive definite functions on locally compact unimodular groups, (to appear), Rend. Sem. Mat. Univ. Padova.

JENSEN MEASURES, SUBHARMONICITY, AND THE CONJUGATION OPERATION

T.W. Gamelin

We consider the notion of "subharmonicity with respect to an algebra of functions", and the dual notion of "Jensen measure". This duality was first exploited effectively by E. Bishop. Subsequently a number of people have contributed towards developing the circle of ideas related to subharjonicity and Jensen measures, including D.A. Edwards, B. Cole, C. Rickart, N. Sibony, A. Debiard and B. Gaveau, and K. Yabuta. A detailed account of these ideas has appeared recently in monograph form [3]. Here I would like to touch upon some of the highlights of this area, referring to [3] for more details and references to the literature.

Subharmonic Functions

There are various levels of abstraction at which one can begin. Let's restrict ourselves immediately to the context of primary concern to us.

Fix a uniform algebra A on a compact space X (a closed subalgebra of C(X) that contains constants and separates points of X), and let M_A denote the maximal ideal space of A. We regard X as a closed subset of M_A, and we regard A as a closed subalgebra of $C(M_A)$.

The continuous subharmonic functions on a closed subset E of M_A are the uniform limits on E of functions of the form

(1) $\max(b, c_1\log|f_1|,\ldots,c_m\log|f_m|)$,

when $b \in \mathbf{R}$, $c_1,\ldots,c_m > 0$, and $f_1,\ldots,f_m \in A$. The continuous subharmonic functions form a convex cone that contains the constants. Any functions $u \subset \mathrm{Re}(A)$ is subharmonic, as can be seen by writing $u = \log|e^g|$, where $g \in A$ satisfies $\mathrm{Re}(g) = u$.

An upper semi-continuous function u from the compact set E to $[-\infty, +\infty)$ is subharmonic on E if u is a decreasing pointwise limit on E of a net of continuous subharmonic functions. It is easy to see that this definition does not broaden the class of continuous subharmonic functions. In the case $E = M_A$, this definition is equivalent, it turns out, to that introduced by C.E. Rickart [14].

Finally, an upper semi-continuous function u from an open subset U of M_A to $[-\infty, +\infty)$ is <u>locally subharmonic</u> if every point of U has a compact neighborhood on which u is subharmonic.

Let us see what this notion of subharmonicity becomes in various special cases.

Let K be a compact subset of the complex plane, and let R(K) denote the uniform closure in C(K) of the functions that are analytic in a neighborhood of K. The maximal ideal space of R(K) coincides with K itself. In this case, the continuous R(K)-subharmonic functions on K are precisely the uniform limits on K of the functions that are continuous and subharmonic (in the usual sense) in a neighborhood of K [3, p. 40]. Moreover, the locally R(K)-subharmonic functions on K^o are the usual subharmonic functions on K^o, except that they are allowed to be identically $-\infty$ on a number of components of K^o.

Thus, for algebras of analytic functions of one complex variable, the notion of A-subharmonicity (subharmonicity with respect to A) generally reduces to the usual notion of subharmonic function. For algebras of analytic functions of several complex variables, the locally A-subharmonic functions are generally the plurisubharmonic functions. More precisely, a theorem of H. Bremermann shows that if U is an open subset of M_A that is covered by "analytic balls" of complex dimension m, coordinatized by functions in A, then the locally A-subharmonic functions on U are precisely the plurisbharmonic functions on U [3,p.93].

One of the principle results concerning A-subharmonic functions is the following theorem proved in [4].

<u>Localization Theorem</u> [3,p.69]: A locally subharmonic function on M_A is subharmonic.

The proof of the Localization Theorem depends upon the application of Rossi's local maximum modulus principle to uniform algebras on certain subhsets of $M_A \times C$ generated by Hartogs polynomials $\sum_{j=0}^{N} f_j \zeta^j$, with coefficients $f_0, \ldots, f_N$ in A. The duality between subharmonic functions and Jensen measures, which we have yet to discuss, also plays a role.

An example of an immediate application of the Localization Theorem, is the following extension of the theorem of Bremermann cited earlier.

<u>Theorem</u>: Let D be a bounded domain in C^n with smooth boundary, such that $\bar{D}$ is an S_δ-set (intersection of domains of holomorphy). Then any continuous real-valued function on $\bar{D}$ that is plurisubharmonic on D can be approximated uniformly on $\bar{D}$ by functions of the form (1), where $b \in \mathbb{R}$, $c_1, \ldots, c_m > 0$, and $f_1, \ldots, f_m$ are holomorphic in a neighborhood of $\bar{D}$.

<u>Proof</u>: We apply the Localization Theorem to the algebra $H(\bar{D})$ generated by the functions holomorphic in a neighborhood of $\bar{D}$. Since $\bar{D}$ is an S_δ-set, $\bar{D}$ coincides with the maximal ideal space of $H(\bar{D})$.

Let $u \in C_R(D)$ satisfy the hypotheses of the Theorem. Since u is plurisubharmonic on D, u is locally $H(\bar{D})$-subharmonic there, by the theorem of Bremermann cited above. Let $p \in \partial D$, let ν be the outer unit normal to ∂D at p, and let B be a small closed ball in C^n centered at p. For $\varepsilon > 0$ small, the translates $u_\varepsilon(z) = u(z - \varepsilon\nu)$ are plurisubharmonic in a neighborhood of $B \cap \bar{D}$. By Bremermann's Theorem again, each u_ε is $H(\bar{D})$-subharmonic on $B \cap D$, and hence so is the uniform limit u. This u is locally $H(\bar{D})$-subharmonic. By the Localization Theorem, u is $H(\bar{D})$-subharmonic, and in fact u can be approximated uniformly on $\bar{D}$ by functions of the form (1), where $f_1, \ldots, f_m$ are chosen from the dense subalgebra of functions holomorphic in a neighborhood of $\bar{D}$.

Jensen Measures and Jensen Boundary

A Jensen measure for a point $\varphi \in M_A$ is a probability measure σ on M_A such that

$$(2) \quad |\log f(\varphi)| \leq \int \log|f| d\sigma, \qquad f \in A.$$

For example, the point mass at φ is always a Jensen measure for φ, and if this is the only Jensen measure, φ is said to belong to the Jensen boundary of A.

As another example, the normalized arc-length measure $d\theta/2\pi$ on the circle $\{|z| = 1\}$ is a Jensen measure for the disc algebra $A(\Delta)$ of functions analytic on the open unit disc $\Delta = \{|z| < 1\}$ and continuous on the closure. In this case, the estimate (2) merely reflects the subharmonicity of $\log|f|$, for f analytic. I do not know when this estimate was first written down. The earliest reference to the related Jensen formula [7] seems to be found in a paper of Jacobi [6], appearing in 1827 in the second volume of Crelle's Journal. In that paper, Jacobi establishes the Jensen formula for a certain class of rational functions, though his proof is valid for entire functions. E. Landau observed in [12, pp. 115-116] that the Jensen formula goes back to Jacobi, and Landau refers to the formula as the "Jacobi-Jensenschen Satz" in his book on ideal theory. Unaware of Landau's remarks, the Ceck mathematician G. Kowalewski [10,11] also cites the Jacobi paper as the first appearance of Jensen's formula. Meanwhile, J. Hadamard [5,p. 50] apparently believed that he was first to discover the Jensen formula. He felt that his reluctance to publish the formula without further significant applications led to his being scooped by Jensen.

Thus it seems appropriate to refer to (2) as the Jensen-Hartogs inequality. In view of the definition of subharmonic functions, the inequality (2) is equivalent to

$$(3) \quad u(\varphi) \leq \int u d\sigma, \quad \text{all subharmonic } u \text{ on } M_A.$$

The Jensen-Hartogs inequality actually characterizes subharmonic functions. Indeed, an upper semi-continuous function u on a compact subset E of M_A is subharmonic if and only if $u(\varphi) \leq \int u d\sigma$ for all $\varphi \in E$ and all Jensen measures σ for φ carried by E. This is a consequence of a theorem of D. A. Edwards [2], which provides the link between subharmonic functions and Jensen measures. Edwards' theorem is related to the Hahn-Banach Theorem. It states that if w is a lower semi-continuous function on a compact subset E of M_A, and $\varphi \in E$, then

$$\sup\{u(\varphi): \quad u \text{ subharmonic on } E,\ u \leq w\}$$

coincides with

$$\inf\{\int w d\sigma: \ \sigma \text{ a Jensen measure on } E \text{ for } \varphi\}$$

The main goal of Edwards' work is to extend the Choquet theory to a context that includes Jensen measures. Edwards shows, for instance, that if A is separable, then the Jensen boundary is a G_δ - set, and every $\varphi \in M_A$ has a Jensen measure that lives on the Jensen boundary.

The Jensen boundary points can be characterized in terms of barriers. A <u>barrier</u> at a point $\varphi \in M_A$ is a subharmonic function u on M_A such that $u(\varphi) = 0$, while $u < 0$ on $M_A \setminus \{\varphi\}$. From the estimate (3), it is clear that φ is a Jensen boundary point, just as soon as there is a barrier at φ. Conversely, it is proved in [4] that if A is separable (for simplicity), and if φ is a Jensen boundary point, then there is a continuous subharmonic function on M_A that is a barrier at φ. This can be regarded as an abstract version of a theorem of M.V.Keldysh, asserting that there is a continuous subharmonic barrier at any regular boundary point of a domain in R^n.

On the basis of the characterization of Jensen boundary points in terms of barriers, and the Localization Theorem, it is shown in [4] that the notion of Jensen boundary point is local. To state the theorem formally, let A_E denote the uniform closure in C(E) of the restriction algebra $A|_E$, whenever E is a closed subset of M_A.

Theorem: [3, p. 72]: Let $\varphi \in M_A$. Suppose there is a compact neighborhood N of φ in M_A such that φ is a Jensen boundary point for A_N. Then φ is a Jensen boundary point for A.

For the metrizable case, the proof proceeds as follows. Since φ is a Jensen boundary point for A_N, there is an A_N-subharmonic function v on N such that $v(\varphi) = 0$, and $v < 0$ on $N\setminus\{\varphi\}$. Choose $\delta > 0$ so that $v \leq -2\delta < 0$ on ∂N. Define a function u on M_A so that $u = -\delta$ on $M_A\setminus N$, while $u = \max(-\delta, v)$ on N. Being the maximum of two A-subharmonic functions, u is clearly A-subharmonic in the interior of N. On the other hand, u is constant, hence A-subharmonic, on a neighborhood of $M_A\setminus N^o$. By the Localization Theorem, u is a barrier at φ, so φ is a Jensen boundary point for A.

For the algebra R(K), K a compact subset of C, the Jensen boundary points are the "stable" boundary points of K, that is, the regular boundary points for the "outer" Dirichlet problem. These are precisely the points p for which the Wiener series

$$(4) \qquad \sum_{k=1}^{\infty} k/\log|1/\mathrm{cap}(E_k)|$$

diverges, where "cap" denotes logarithmic capacity, and E_k is the annulus $\{2^{-k} \leq |z-p| \leq 2^{-k+1}\}$. The fact that the Jensen boundary is local corresponds to the fact that the divergence of the series (4) depends only on the tail of the series.

There is no known characterization of the Jensen boundary points for the algebra $H(\bar{D})$, when D is a pseudoconvex domain in C^n with smooth boundary. It is known, though, that if every point of ∂D with one possible exception p is a strictly pseudoconvex boundary point, then p is a Jensen boundary point. Under the same hypotheses, it is not known however whether p is necessarily a peak point.

Some Directions for Study

Equipped with a notion of subharmonicity, one can study an abstract version of Bremermann's Dirichlet problem. Let U be an open subset of $M_A \setminus X$, and let $w \in C_R(\partial U)$. Consider the family of continuous locally subharmonic functions u on U such that

$$\limsup_{U \ni \varphi \to p} u(\varphi) \leq w(p)$$

for all $p \in \partial U$. The upper envelope of all such functions is denotebly $\tilde{w}$, and called the solution to Bremermann's generalized Dirichlet problems with boundary data w. It turns out that $\tilde{w}$ is a locally "quasi-subharmonic" function on U. Under certain conditions, including regularity of the boundary of U, the solution w is a continuous locally subharmonic function on U that attains the boundary values w continuously at ∂U.

The functions that arise as solutions to this Dirichlet problem form by themselves an interesting object for study. They can be defined locally, and they are called Bremermann functions. In one complex variable, the Bremermann functions are simply the harmonic functions. In several complex variables, the smooth Bremermann functions are the smooth plurisubharmonic functions u that satisfy

$$(5) \quad \det\left(\frac{\partial^2 u}{\partial z_j \partial z_k} \right) = 0.$$

In this case, Bremermann's generalized Dirichlet problems boils down to finding a solution of the Monge-Ampère equation (5) with prescribed boundary data.

Another direction which is natural for investigation involves the notion of a set of capacity zero. Bishop introduced such a notion in a rather abstract context in [1], where he proved a generalization of the Remmert-Stein Theorem on removable singularities of analytic sets.

The notion of set of zero capacity introduced in [4] is virtually the same as that of Bishop. Let U be an open subset of $M_A \setminus X$. A subset E of U is of zero capacity relative to U if there is a subharmonic function u on U such that $u = -\infty$ on E, while the set $\{u = -\infty\}$ has no interior. Any definition in terms of subharmonic functions has a dual in terms of Jensen measures. In the case at hand, we let E^* denote the set of all $\varphi \in U$ that have a Jensen measure with positive mass on E. Then E has zero capacity if and only if E^* has no interior.

For the algebra R(K), K a compact subset of C, a relatively closed subset of K^o has zero capacity in this sense if and only if it has zero logarithmic capacity. For subsets of C^n, the sets of zero capacity correspond roughly to the pluripolar sets, defined formally as follows. A subset E of a domain D in C^n is a pluripolar set if there is a plurisubharmonic function v on D, not identically $-\infty$, such that $v = -\infty$ on E.

Recently B. Josefson [8] has established that the notion of "pluripolar set" is local; if E is a subset of D, and if every $p \in E$ is included in an open ball $B \subset D$ such that $E \cap B$ is a pluripolar subset of B, then E is a pluripolar subset of D. It is tempting to conjecture that the analogous statement is valid for uniform algebras, that is, that the notion of "set of zero capacity" is local. Unfortunately Josefson's proof does not seem to generalize, and in fact his estimates depend very much on the dimension of the ambient complex space. In this connection, it would be of interest to find a simpler proof of Josefson's theorem.

Estimates for Conjugate Functions

There is another area in which the duality between subharmonic functions and Jensen measures plays a crucial role. Fix a continuous real-valued function H on the complex plane C, and consider the problem of determining when

(6) $\int H(u(e^{i\theta}), {}^*u(e^{i\theta}))d\theta \geq 0$

for all trigonometric polynomials u. Here *u is the conjugate trigonometric polynomial of u, determined so that $u + i{}^*u$ is analytic and ${}^*u(0) = 0$. When $H = c|x|^p - |y|^p$ we are dealing with the M. Riesz estimate

(7) $\int |{}^*u|^p \, d\theta \leq c\int |u|^p d\theta.$

The Zygmund estimate

$$\int |{}^*u \, d\theta \leq \alpha \int |u|\log^+|u|d\theta + \beta$$

also falls in this context. If we allow H to be discontinuous, we may also treat weak-type estimates as a special case of (6).

There is a standard technique for obtaining the estimate (6), under the condition

(8) There exists a subharmonic function h on C such that $h \leq H$, with $h \geq 0$ on $\mathbb{R}$.

When this condition is valid, we merely note that $h(u,{}^*u)$ is also subharmonic, so that the inequalities

(9) $0 \leq h(u(0),0) \leq \int h(u, u)\frac{d\theta}{2\pi} \leq \int h(u,{}^*u)\frac{d\theta}{2\pi}$

yield (6). To illustrate this procedure, consider the proof of the M. Riesz estimate given by S.K. Pichorides [13]. It is a modification of an earlier proof of A.P. Calderón, and it leads to sharp constants for the M. Riesz estimate.

Assume $1 < p < 2$. Choose $\alpha > 0$ and $\gamma > 0$ so that

(10) $\alpha\cos(p\theta) \leq \gamma\cos\theta|^p - |\sin\theta|^p, \; -\frac{\pi}{2} \leq \theta \leq \frac{\pi}{2}$.

To do this, one chooses $\alpha > 0$ large so that (10) is valid near $|\theta| = \pi/2$, and then γ large so that (10) is valid elsewhere. Now define h harmonic on the right half plane, in polar coordinates, by

$$h = \alpha r^p \cos(p\theta), \qquad r > 0, \ -\frac{\pi}{2} \leq \theta \leq \frac{\pi}{2},$$

and extend h to the left half plane so as to be symmetric about the imaginary axis. One checks that locally at each point of the positive or negative imaginary semi-axis, h is the maximum of two harmonic functions. Hence h is subharmonic. The estimate (10) becomes

$$h(x,y) \leq \gamma |x|^p - |y|^p.$$

The calculation (9) becomes

$$0 \leq 2\pi h(u(0),0) \leq \int h(u,{*u})\,d\theta \leq \gamma \int |u|^p d\theta - \int |{*u}|^p d\theta,$$

and this proves the M. Riesz theorem, at least for $1 < p < 2$.

The other classical estimates for trigonometric polynomials can be proved in much the same way. Each estimate requires an inspired choice of h. B. Cole has clarified this situation, by proving in effect that all valid integral estimates can be proved in this way. More specifically, he proves that (6) is valid for all trigonometric polynomials u if and only if (8) is valid [3, p. 133].

Cole's proof gives further information not contained in the statement of this theorem. It shows that if the estimate (6) fails for some trigonometric polynomial u, then it fails for functions of a very special form. Often it fails for conformal maps $u + i{*u}$ of the open unit disc onto a domain in the plane that can be devined from H. Following this line of reasoning, Cole had obtained independently the best possible constants in the M. Riesz estimate.

Where do Jensen measures enter ? They enter in Cole's proof. They also enter into a second part of Cole's theorem, which we proceed to discuss.

Consider again the uniform aglebra A, and fix $\varphi \in M_A$. For $u \in \mathrm{Re}(A)$, define ${*u} \in \mathrm{Re}(A)$ so that $u + i{*u} \in A$ and ${*u}(\varphi) = 0$. Unfortunately u is not determined uniquely by u. However, *u is determined uniquely on the closed support of any representing measure for φ. Recall that a representative measure for φ is a probability measure τ such that

$$(11) \quad f(\varphi) = \int f d\tau, \qquad \text{all } f \in A.$$

If $v \in \mathrm{Re}(A)$ also satisfies $u + iv \in A$ and $v(\varphi) = 0$, then $*u - v \in A$ vanishes at φ. Applying (11) to $f = (*u-v)^2 \in A$, we obtain

$$0 = \int (*u-v)^2 d\tau,$$

and indeed $v = *u$ on the closed support of τ.

The additional statement of Cole's theorem is that if H satisfies (6) for all trigonometric polynomials u, then in fact

$$(12) \quad \int H(u,*u)d \geq 0$$

for all uniform algebras A, all $\varphi \in M_A$, all $u \in \mathrm{Re}(A)$, and all Jensen measures σ for φ. To prove (12), one notes that if $u \in \mathrm{Re}(A)$, and if h is subharmonic on C, then $h(u,*u)$ is A-subharmonic. Thus, using the Jensen-Hartogs inequality we obtain

$$0 \leq h(u(\varphi),0) \leq \int h(u,*u)d\sigma \leq \int H(u,*u)d\sigma,$$

just as in (9).

In particular, the M. Riesz estimate is valid for any Jensen measure, with the same constants as are valid for trigonometric polynomials. As an application, consider a bounded domain D in C^n with smooth boundary, let dS be the surface measure on ∂D, and fix $\zeta \in D$. E.L. Stout [15] has established the M. Riesz estimate

$$\int |*u|^p dS \leq c \int |u|^p dS,$$

valid for $u + i*u$ analytic on D and continuous on $\bar{D}$, and $u(\zeta) = 0$. We see now that this estimate follows from Cole's theorem, once we observe that the harmonic measure for p on ∂D, which is a Jensen measure for p, is mutually boundedly absolutely continuous with respect to the area measure dS.

In closing, let me discuss one of the versions of a related theorem due to Cole, involving representing measures rather than Jensen measures.

Theorem [3, p. 109]: Let H be a continuous real-valued function on an open subset D of the complex plane, and let $p \in D$. Then the following are equivalent.

(i) $\int H \cdot f d\tau \geq 0$ for all uniform algebras A, all $\varphi \in M_A$, all representing measures τ for φ, and all $f \in A$ satisfying $f(M_A) \subset D$ and $f(\varphi) = p$.

(ii) There is an analytic function g on D such that $\mathrm{Re}(g) \leq H$, while $g(p) \geq 0$.

Again the existence of the auxiliary function g yields the estimate (i) rather trivially:

$$0 \leq g(p) = g(f(\varphi)) = \int g \cdot f d\tau \leq \int H \cdot f d\tau.$$

The proof that all such estimates can be obtained by the use of an analytic auxiliary function requires some effort.

Thus, proofs of classical estimates involving analytic auxiliary functions extend to arbitrary representing measures, while those involving subharmonic functions extend to Jensen measures. It turns out that the M. Riesz estimates (7) simply fail for arbitrary representing measures, unless p is an even integer. On the other hand, the classical proof of M. Riesz, which depends on an analytic auxiliary function, shows that the estimate (7) is valid provided that $1 < p < \infty$ is not an odd integer, and provided that $u > 0$, that is, the range of $f = u + i{*}u$ lies in the right half-plane.

Why does the Riesz proof fail at the odd integers ? An answer has been discovered recently by H. König [9] and K. Yabuta. They have produced examples of uniform algebras and representing measures for which the M. Riesz estimate simply fails at the odd integers, even for functions with values in the right half-plane. Thus, the M. Riesz proof, using an analytic auxiliary function as it does, cannot possibly cover the case in which p is an odd integer.

References

[1]. Bishop, E., Conditions for analyticity of certain sets, Mich.Math. Journal 11 (1964) 289-304.

[2]. Edwards, D.A., Choquet boundary theory for certain spaces of lower semi-continuous functions, in Function Algebras, F. Birtel (ed.), Scott, Foresman and Co. (1966), 300-309.

[3]. Gamelin, T.W., Uniform Algebras and Jensen Measures, London Math. Soc. Lecture Notes Series, No. 32, Cambridge University Press, 1978.

[4]. Gamelin, T.W., and Sibony, N., Subharmonicity for uniform algebras, J. Functional Analysis, to appear (1979).

[5]. Hadamard, J., The Psychology of Invention in the Mathematical Field, Princeton University Press, 1949.

[6]. Jacobi, C.G.J., Über den Ausdruck der verschiedenen Wurzeln einer Gleichung durch bestimmte Integrale, Journal für die Reine und Angewandte Math. 2 (1827) 1-8.

[7]. Jensen, J.L.W.V., Sur un nouvel et important théorême de la théorie des fonctions, Acta Math. 22 (1899) 359-364.

[8]. Josefson, B., On the equivalence between locally polar and globally polar sets for plurisubharmonic functions on C^n, Arkiv für Mat. 16 (1978) 109-115.

[9]. König, H., On the Marcel Riesz estimation for conjugate functions in the abstract Hardy Theory, to appear.

[10]. Kowalewski, G., Ein Funktionentheoretischer Satz Jacobis, Jahresbericht der Deutschen Mat. Ver. 27 (1918) 53-55.

[11]. Kowalewski, G., Bemerkung zu meinem Aufsatz über einen funktionentheoretischen Satz Jacobis, Jahresbericht der Deutschen Mat. Ver. 27 (1918) p. 160.

[12]. Landau, E., Über eine Aufgabe aus der Funktionentheorie, Tohoku Math. J. 5 (1914) 97-116.

[13]. Pichorides, S.K., On the best values of the constants in the theorems of M. Riesz, Zygmund and Kolmogorov, Studia Math. 44 (1972) 165-179.

[14]. Rickart, C.E., Plurisubharmonic functions and convexity properties for general function algebras, Trans. A.M.S. 169 (1972) 1-24.

[15]. Stout, E.L., H^p functions on strictly pseudo-convex domains, Amer. J. Math. 98 (1976) 821-852.

TWO CONSTRUCTIONS IN BMO

John B. Garnett

We discuss two theorems on functions of bounded mean oscillation, each proved by geometric construction.

§1. Norm BMO in the usual way,

$$\|\phi\|_* = \sup_Q \frac{1}{|Q|} \int |\phi-\phi_Q|dx$$

where Q is a cube in R^d, $|Q|$ is its measure, and ϕ_Q is the average of ϕ over Q. The first theorem estimates the distance

$$\inf_{g\varepsilon L} \|\phi-g\|_*$$

by the constant in the John-Nirenberg inequality

$$(1) \qquad \sup \frac{|\{x \varepsilon Q : |\phi(x)-\phi_Q| > \lambda\}|}{|Q|} < e^{-A\lambda}$$

for λ large. Define

$$A(\phi) = \sup\{A : (1) \text{ holds for all } \lambda \geq \lambda_0(A,\phi)\}.$$

By the John-Nirenberg theorem [10], $\phi \varepsilon \mathrm{BMO}$ if and only if $A(\phi) > 0$, and in fact

$$A(\phi) \geq C/\|\phi\|_*$$

It is clear that

$$A(\phi) = +\infty \quad \text{if } \phi \varepsilon L^\infty$$

and that

$$A(\log|x|) = 1.$$

THEOREM 1: <u>There are constants</u> c_1 <u>and</u> c_2, <u>depending only on the dimension, such that</u>

$$c_1/A(\phi) \leq \inf_{g\in L^\infty} \|\phi - g\|_* \leq c_2/A(\phi).$$

The left inequality is trivial, and so the right inequality is the real theorem, proved jointly with my student P. Jones [7].

By Fefferman's theorem [6],

$$\text{(2)} \qquad \phi = g_0 + \sum_{j=1}^{d} R_j g_j$$

where R_j is the Riesz transform, $\widehat{R_j g} = \frac{-x_j}{|x|}\hat{g}$, and where $g_0, g_1, \ldots, g_d \in L^\infty$ with

$$\|\phi\|_* \sim \sum_0^d \|g_j\|_\infty .$$

Thus Theorem 1 can be reformulated as

$$1/A(\phi) \sim \inf\{\sum_1^d \|g_j\|_\infty \ : \ (2) \text{ holds for some } g_0 \in L^\infty\}$$

A consequence is a higher dimensional Helson-Szegö theorem. Let $w \geq 0$ be a weight function on R^d. The Riesz transforms are bounded on $L^2(wdx)$ if and only if

$$(A_2) \qquad \sup_Q \left(\frac{1}{|Q|}\int_Q w\,dx\right)\left(\frac{1}{|Q|}\int_R\!\int_Q \frac{1}{w}\,dx\right) < \infty$$

See Hunt, Muckenhoupt and Wheeden |9| and Coifman and Fefferman [2]. When $d = 1$ there is an earlier necessary and sufficient condition, due to Helson and Szegö

$$\log w = g_0 + Hg_1, \quad \| g_1 \|_\infty < \pi/2, \quad g_0 \in L^\infty,$$

where H denotes the Hilbert transform or conjugate function. These two results can be connected to Theorem 1 via

<u>LEMMA 1</u>: <u>If</u> $= \log w$, <u>then</u>

$$A(\phi) > 1 \leftrightarrow w \text{ has } (A_2).$$

This lemma is easily derived from Jensen's inequality and the fact that (A_2) implies (A_p) for some $p < 2$. For $d = 1$ the lemma means that Theorem 1 follows from the equivalence between (A_2) and the Helson-Szegö condition. See [9] and [5]. For $d > 1$ we obtain a new Helson-Szegö result.

<u>COROLLARY 1</u>: <u>There are constants</u> $B_1(d)$ <u>and</u> $B_2(d)$ <u>such that if</u> (2) <u>holds with</u> $\sum_1^d \|g_j\|_\infty < B_1$, <u>then</u> e^ϕ <u>satisfies</u> (A_2). <u>Conversely, if</u> e^ϕ <u>has</u> (A_2), <u>then</u> (2) <u>holds for</u> ϕ <u>with</u> $\sum_1^d \|g_j\|_\infty < B_2$.

My student P. Jones has observed that $B_1 \neq B_2$ in the case $d > 1$, no matter how we norm the vector $(\|g_1\|_\infty, \|g_2\|_\infty, \ldots, \|g_d\|_\infty)$. Perhaps we must accept Lemma 1 as the sharp higher dimensional Helson-Szegö theorem.

§2. To give a rough (and slightly inaccurate) proof of Theorem 1, take $d = 1$. Let $A < A(\phi)$ and fix $\lambda > \lambda_0(A, \phi)$ $\lambda > 10\|\phi\|_*$. We use (1) to build a function f such that

$$|\phi - f| < C\lambda$$

$$\|f\|_* \leq C/A$$

The only interesting case is when $1/A$ is small compared to $\|\phi\|_*$.

The function f is a sum of little functions called adapted functions. Let I be a dyadic interval with triple $\tilde{I}$. Say that $a(x)$ is adapted to I if $a(x)$ is Lipschitz and

(a) $|a(x)| \leq 1$

(b) $a(x) = 0, \ x \in \tilde{I}$

(c) $|\nabla a(x)| \leq 1/|I|$, a.e.

<u>LEMMA</u> 2: <u>If</u> I_j <u>is a sequence of dyadic intervals such that</u>

$$\sum_{I_j \subset I} |I_j|/|I| \leq K \tag{3}$$

<u>for every interval</u> I, <u>and if</u> $a_j(x)$ <u>is adapted to</u> I_j, <u>then</u>

$$\|\sum a_j\|_* \leq cK.$$

The proof is not difficult. Use (c) to control $\sum_{I_j \subset I} a_j(x)$ on an interval I.

Change scale so that $A = \log 4$. Fix a dyadic interval I and suppose $\phi_I = 0$. Let λ be an integer, $\lambda > \lambda_0(A,\phi)$, $\lambda > 10\|\phi\|_*$. Let $R = \{R_1, R_2, \ldots\}$ consist of the maximal dyadic subintervals of I for which $\phi_{R_K} > \lambda$. By (1),

$$\sum |R_k|/|I| < c\, 4^{-\lambda}$$

and we can select dyadic intervals $\{I_j\}$ contained in I such that (3) holds with fixed K and such that each R_k is covered by λ intervals I_j. This means there are functions $a_j(x)$ adapted to I_j so that $f_1 = \sum a_j$ satisfies

$$0 \leq f_1 \leq \lambda$$

$$f_1 = \lambda \quad \text{on } R_k$$

$$\|f_1\|_* \leq C$$

and f_1 has support in $\tilde{I}$.

Repeating this process with $\phi-\lambda$ on each R_k and continuing, we obtain $f_1, f_2, \ldots$ so that

$$\phi-(f_1 + f_2+\ldots) < C\lambda$$

and

$$\|f_1 + f_2 +\ldots\|_* \leq C.$$

Some technical difficulties involving large positive and negative values of $\phi(x)$ must be confronted to get the full result

$$-C\lambda < \phi -(f_1 + f_2 + \ldots) < C\lambda$$

and we refer to [7] for the rest of the story.

Adapted functions can be eliminated from the proof by using the fact that

$$\|\log M(\chi_E)\|_* \leq C$$

where M is the Hardy-Littlewood maximal function. See [4] or [3]. Setting $E = \cup R_k$, one can then simply take

$$f_1 = (\alpha + \beta \log M(\chi_E))^+$$

in the above argument. See [11], where this method is used to obtain Theorem 1 on spaces of homogeneous type. However, the basic difficulties concerning changes of sign remain the same, and adapted functions do provide certain conceptual advantages.

The construction is similar to Carleson's proof [1] that

(4) $$\phi(t) = \int \frac{1}{y} K\left(\frac{x-t}{y}\right) d\mu(x,y) + \text{bounded}$$

where K is a kernel like the Poisson kernel and where μ is a Carleson measure:

(5) $$|\mu|(I \times (0,|I|)) \leq M|I|.$$

Condition (3) is suggestive of Carleson measures and the proof of Theorem 1 amounts to obtaining (4) with the constant M in (5) as small as possible.

§3. To state the second theorem we keep $d = 1$, although the result holds for $d > 1$. Let $u(x,y)$ be the Poisson integral of a BMO function ϕ. Then $y|\nabla u|^2 dxdy$ is a Carleson measure

(6) $$\iint_{I\times(0,|I|)} y|\nabla u|^2 dxdy \leq C|I|.$$

For some applications it would be more desirable if $|\nabla u| dxdy$ were a Carleson measure. That is unfortunately false [12]. As a compromise we have

THEOREM 2: *If* $\phi \in BMO$ *and if* $\varepsilon > 0$, *then there is* $\psi(x,y) \in C^\infty(y > 0)$ *such that*

(7) $$|u(x,y) - \psi(x,y)| < \varepsilon$$

and

(8) $$\iint_{I\times(0,|I|)} |\nabla u| dxdy \leq (C(\varepsilon, \|\phi\|_*))\,|I|$$

for all intervals I.

Here is an application of Theorem 2 not an obvious direct consequence of (6). Fix $h > 0$ and let $N_\varepsilon^h(x)$ be the number of ε-oscillations of $u(x,y)$ on the segment $(0,h]$. That is, $N_\varepsilon^h u(x) > n$ if there are $0 < y_0 < y_1 < \dots < y_n \le h$ such that

$$|u(x,y_{j+1}) - u(x,y_j)| \ge \varepsilon.$$

By Fatou's theorem $N^h(x) < \infty$ almost everywhere. By Theorem 2,

$$\frac{1}{h} \int_{x_0}^{x_0+h} N_\varepsilon^h u(x)\,dx \le C'(\varepsilon, \|\phi\|_*)$$

for every x_0.

Theorem 2 was suggested by work of Varopoulos [13], [14]. The construction resembles that used in the Corona Theorem, which in turn can be derived from Theorem 2. The function $\psi(x,y)$ cannot in general be harmonic.

We briefly describe the proof. By (6) we can take $\psi = u$ when $y|\nabla u|$ is large. On the place where $y|\nabla u|$ is small it is possible to compare $u(x,y)$ to a dyadic martingale, by replacing $u(x,y)$ with its average over certain horizontal slits. A simple stopping time procedure then yields a piecewise constant solution of (7) and (8) on $\{y|\nabla u|$ small$\}$. It is not difficult to mollify this solution into a C^∞ function. The details will appear elsewhere.

REFERENCES

[1]. L. Carleson, "Two Remarks on H^1 and BMO", Advances in Math. 22(1976) 269-277.

[2]. R.R. Coifman and C. Fefferman, "Weighted Norm Inequalities for Maximal Functions and Singular Integrals," Studia Math. 51 (1974) 241-250.

[3]. R.R. Coifman and R. Rochberg, "Another Characterization of BMO," preprint.

[4]. A.Cordoba and C. Fefferman, "Weighted Norm Inequalities for the Hilbert Transform," Studia Math. 62, (1976) pp. 97-101.

[5]. C. Fefferman, "Recent Progress in Classical Fourier Analysis," Proc. Int. Cong. Math. Vancouver, 1974, I, pp. 95-118.

[6]. C. Fefferman and E.M. Stein, "H^p Spaces of Several Variables," Acta Math. 129 (1972) pp. 137-193.

[7]. J.B. Garnett and P.W. Jones, "The Distance in BMO to L^∞", Annals of Math. (to appear).

[8]. H. Helson and G. Szegö, "A Problem in Prediction Theory", Ann. Math. Pure Appl. 51 (1960) pp. 107-138.

[9]. R.A. Hunt, B. Mickenhoupt and R.L. Wheeden, "Weighted Norm Inequalities for the Conjugate Function and Hilbert Transform," Trans. Amer. Math. Soc. 176 (1973), pp. 227-251.

[10]. F. John and L. Nirenberg, "On Functions of Bounded Norm Oscillations," Comm. Pure Appl. Math. 14 (1961) pp. 415-426.

[11]. P.W. Jones, "Constructions with Functions of Bounded Mean Oscillations," Thesis UCLA 1978.

[12]. W. Rudin, "The Radial Variation of Analytic Functions," Duke Math. J 22 (1955) pp. 235-242.

[13]. N.Th. Varopoulos, "A Remark on BMO and Bounded Harmonic Functions," Pac. Jour. Math. 72 (1978).

[14]. N.Th. Varopoulos, "BMO Functions and the -equation," Pac.Jour. Math. 71 (1977) pp. 221-273.

University of California, Los Angeles

University de Paris-Sud, Orsay

Maximal function characterization

of H^p for the bidisc.[1)]

R.F. Gundy, Rutgers University

New Brunswick, New Jersey, 08903 USA

In this note, we describe some results, obtained in collaboration with E.M. Stein, which generalize the maximal function characterization of H^p obtained on [2]. An announcement of our results appears in [5], and a detailed discussion of the proofs is given in [4]. Here we shall give a statement of the main theorem and a sketch of the proof in a simplified context.

Let $F(z_1,z_2)$ be a holomorphic function in $D=\{(z_1,z_2):|z_i|<1,\ i=1,2\}$. Such a function is said to belong to the Hardy class H^p, $0<p<\infty$, if

$$(1) \quad \sup_{r_1,r_2<1} \int_0^{2\pi}\int_0^{2\pi} |F(r_1e^{i\theta_1}, r_2e^{i\theta_2})|^p \, d\theta_1 d\theta_2 < \infty$$

Given such a function F, we may express it as the Poisson integral of a distribution supported on the "distinguished boundary" $\partial D_1 \times \partial D_2$. That is, F arises as a convolution of a distribution f with the product kernel $(P_1+i\tilde{P}_1)(P_2+i\tilde{P}_2)$, where P_j is the one dimensional Poisson kernel, and $(\tilde{P}_i$ its conjugate kernel. Here, however, we see the first departure from the one-dimensional setting: Given F, we can construct three associated functions, F_k, $k=1,2,3$. The functions F and F_k, $k=1,2,3$ are obtained by convolving $(P_1 \pm i\tilde{P}_1)(P_2 \pm i\tilde{P}_2)$ with the distribution f for each of the various choices signs. Each function F_k is holomorphic in an appropriate domain. Thus,

$F_1=(P_1-iP_1)(P_2+iP_2)$ will be holomorphic in $(\bar{z}_1,z_2)$, F_2 in

1) This research was in part supported by NSF Grant MCS 78-15273.

$(z_1,\bar{z}_2)$, and F_3 in $(z_1,\bar{z}_2)$. Moreover, their H^p norms are defined by (1), and it is clear that

$$\|F\|_p = \|F_k\|_p,\ k=1,2,3.$$

From the fact that a single function F gives rise to the system F, F_k, k = 1,2,3 in the manner just described, we see that the notion of "conjugate function" is more complicated that in the one dimensional setting. There, we may write a holomorphic function in the form $F = u + i\tilde{u}$, where $\tilde{u}$ is the conjugate of the harmonic function u. This conjugate function is uniquely determined up to the constant $\tilde{u}(0)$, which we take to be zero. In the bidisc, however, we see that <u>given a single real, biharmonic function</u> $u(z_1,z_2)$ (harmonic in each variable separately) we may associate to it three conjugates, u_1, u_2, u_{12} where u_1 is the conjugate of u in the first variable, u_2 in the second, and u_{12} in both variables.

We seek a characterization of H^p in terms of maximal functions of the following sort:

The <u>nontangential maximal function</u>.

$$N(u)(\theta_1,\theta_2) = \sup_{z\,\varepsilon\,\Gamma(\theta)} |u(z)|$$

where $z = (z_1,z_2)$ and $\Gamma(\theta)$ is the product of two cones $\Gamma(\theta_1) \times \Gamma(\theta_2)$, where $\Gamma(\theta_j)$ is a cone contained in D with apex at $\exp(i\,\theta_j)$, and fixed opening angle.

The <u>Brownian maximal function</u>

$$u^* = \sup_{t,s} |u(z_t,z_s)|$$

where z_t, and z_s are two independent two dimensional Brownian motions with <u>independent</u> time parameters t,s.

The main result is the following.

Theorem. _Let_ u _be_ _a_ _biharmonic_ _function_ _defined_ _for_ (z_1,z_2) _belonging_ _to_ D. _If_ N(u) _belongs_ _to_ L^p _for_ _some_ p, $0 < p < \infty$, _then_ _it_ _follows_ _that_ $N(u_1)$, $N(u_2)$, _and_ $N(u_{12})$ _belong_ _to_ L^p, _so_ _that_ _the_ _corresponding_ _functions_ F, F_k, $k = 1,2,3$ _each_ _belong_ _to_ H^p. _Conversely_, _if_ F _belongs_ _to_ H^p, _then_ _the_ _four_ _corresponding_ _real_ _biharmonic_ _functions_ _are_ _such_ _that_ $N(u_i)$ _belongs_ _to_ L^p, $i = 0,1,2,12$.

The proof of this theorem is a combination of geometry and probability theory. Here are some of the ideas:

(1) The connection between the given biharmonic function u and its conjugates u_1, u_2 and u_{12} is an appropriate "area" integral A(u). This is a two-parameter generalization of the classical Lusin area integral; the integrand is the modulus of an interated gradient $|\nabla_1 \nabla_2 u(z_1,z_2)|$. As in the one-dimensional case, this is the object that is stable under conjugation.

(2) There are Brownian motion analogues of N(u) and the area integral. These too may be used to characterize H^p. In fact, the difference between the one and two dimensional problems is most clearly seen when we consider the Brownian motion analogue of N(u). We define $u^* = \sup_{t,s} |u(z_t,z_s)|$ where z_t and z_s are two independent two dimensional Brownian motions, moving with independent time parameters t,s in each disc D_1 and D_2. The process $u(z_t)$ may be shown to be a martingale with a two-dimensional time parameter. Since the parameter set (t,s) has only a partial order, we seem to be deprived of the principal tool for analyzing martingales: the _stopping_ _time_. Part of the interest of the theorem is the fact that it is possible to surmount this difficulty. Using an unpublished theorem of C. Fefferman, suitably adapted, we show that $\|N(u)\| \leq C_p \|A(u)\|_p$ for $0 < p < \infty$. The proof, however, uses the probability analogues of both of these functions, as well as the equivalence between the geometric and probabilistic approaches. Furthermore, the proof shows that $\{N(u) < \infty\} \supset \{A(u) < \infty\}$, up to a set of measure zero.

(3) In contrast to the above inequality, it's converse $\|A(u)\|_p \leq C_p \|N(u)\|_p$, $0 < p < \infty$, is proved without any appeal to probability. Our proof is a sharpening of some estimates used by M.P. and P. Malliavin [6] to obtain $\{A(u) < \infty\} \supset \{N(u) < \infty\}$, again up to a set of measure zero. Because the calculations are rather tedious in the case of two dimensions, we outline here the basic idea <u>in the context of one dimension</u>.

The inequality we wish to prove relates the area integral

$$A(u)(e^{i\theta}) = \left[\int_{\Gamma(\theta)} |\nabla u|^2 (z)\, dm\right]^{1/2}$$

to the maximal function $N(u)$. (Here, as before, $\Gamma(\theta)$ is a cone centered at $e^{i\theta}$, and dm is two-dimensional Lebesgue measure in D.

The inequality to be obtained is the following

$$(1) \qquad m_1(A(u) > \lambda) \leq C\lambda^{-2} \int |N(u)|^2\, dm_1 + Cm_1(N(u) > \lambda)$$

where m_1 is one dimensional Lebesgue measure. Similar inequalities may be found in [1], (Lemma 2.2) for martingales and in Fefferman and Srein [3], for the area and maximal functions. The Fefferman-Stein strategy for proving (1) is as follows: We let $G = \{\theta : N(u)(e^{i\theta}) \leq \lambda\}$ and $G^+ = \bigcup_{\theta \in G} \Gamma(\theta)$. In the region G^+, we know that $|u| \leq \lambda$. Its boundary ∂G^+ resembles a mountain range based on the circle $|z| = 1$.
Since $m_1(A(u) > \lambda) = m_1(A(u) > \lambda, G) + m_1(A(u) > \lambda, G^c)$ and $m_1(A(u) > \lambda, G^c) \leq m_1(G^c) = m_1(N(u) > \lambda)$, we are left with the estimation of $m_1(A(u) > \lambda, G)$. By Chebychev's inequality,

$$m_1(A(u) > \lambda, G) \leq \lambda^{-2} \int_G A^2(u)\, dm_1$$

The Fefferman-Stein approach to the estimation involves the use of Green's theorem for the region G^+. Thus, by Fubini's theorem

$$(2)\qquad \int_G A^2(u)\,dm_1 \le C \int |\nabla\ |^2 \chi_{G^+}\, g\, dm$$

$$\le c \int_{\partial G^+} |u|^2 dm_1 \text{ (Green's theorem)}$$

$$\le c \int_G |N(u)|^2 dm_1 + C\lambda^2 m_1(N(u) > \lambda$$

Since $|u| \le \lambda$ on ∂G^+ and the arc length

$$m_1(\partial G^+ \cap \{|Z| < 1\}) \le C\, m_1(N(u) > \lambda).$$

Here, and below, $g(z) = \log|z|^{-1}$, the Green's function for D with pole at the origin.

In the bidisc, however, the region G^+ has an extremely complicated boundary; it is not clear how Green's theorem may be applied in this setting. On the other hand, Green's theorem is applicable to any (four dimensional region that is the Cartesian product of two dimensional regions: we may simply iterate the formula for each dimension. In particular, Green's theorem may be applied to the bidisc itself. With this in mind, let us examine inequality (2). Green's theorem is applied to the function $|\nabla u|^2$ integrated over the region G^+ with respect to the measure $\log|z|^{-1}\, dm(z)$. To avoid this awkward region G^+, we replace the function $|\nabla u|^2 = 2^{-1}\Delta u^2$ by the function $\Delta(u^2\tilde{\chi}_{G^+})$, where $\tilde{\chi}_{G^+}$ is a smooth (C^∞) version of χ_{G^+}. The construction of $\tilde{\chi}_{G^+}$ can be carried out in the bidisc as easily as in the disc. Having done this, we can calculate

$$(3)\qquad \Delta(u^2\tilde{\chi}_{G^+}) = (\Delta u^2)\,\tilde{\chi}_{G^+} + \text{error}.$$

Green's theorem applies to the left-hand side of this equation since the integral may be taken over the entire disc (or bidisc). The calculation gives a boundary integral that is estimated as in (2), and we obtain the required bound. The first member on the right side of (3) gives rise to the integral that appears in (2). Therefore, it remains for us to estimate the error term integral. But this is dominated by the integral

$$2 \int |\nabla u^2| |\nabla \tilde{\chi}_{G^+}| g \, dm \leq C \int |u| |\nabla u| |\nabla \tilde{\chi}_{G^+}| g \, dm$$

To estimate this last integral, we make some preliminary remarks. 1) It is possible to construct the smooth approximation $\tilde{\chi}_{G^+}$ of the characteristic function $\tilde{\chi}_{G^+}$ so that (a) $|\nabla \tilde{\chi}_{G^+}| \leq C \tilde{\chi}_{G^+} |\nabla V_G|$ where V_G is the Poisson integral of the boundary set G, 2). Furthermore, by contracting the set G a little at the outset (so that $m_1(N(u) > \lambda) \leq C m_1(G)$) we can arrange matters so that $|u| \leq \lambda$ on the support of $\tilde{\chi}_{G^+}$. With these provisions, the last integral may be estimated using Schwarz's inequality:

$$\int |u| |\nabla u| |\nabla \tilde{\chi}_{G^+}| g \, dm \leq \lambda \int |\nabla u| |\nabla \tilde{\chi}_{G^+}| g \, dm$$

$$\leq c \lambda \left(\int |\nabla u|^2 \tilde{\chi}_{G^+} g \, dm \right)^{1/2} \left(\int |\nabla V_G|^2 g \, dm \right)^{1/2}$$

$$\leq c [\lambda^2 m(N(u) > \lambda)]^{1/2} \left[\int |\nabla u|^2 \tilde{\chi}_{G^+} g \, dm \right]^{1/2}$$

The last inequality is obtained by using Green's theorem and the fact that $|\nabla(1 - V_G)| = |\nabla V_G|$, so that

$$\int |\nabla V_G|^2 g \, dm \leq c \, m(G^c)$$

$$= c \, m(N(u) > \lambda).$$

Thus, we have shown that the error term integral is bounded by the above product. This product is, in turn, bounded:

$$[\lambda^2 m(N(u) > \lambda)]^{1/2} | \int |\nabla u|^2 \chi_{G^+} g \, dm]^{1/2}$$

$$\leq \delta^{-1}[\lambda^2 m(N(u) > \lambda)] + \delta \, [\int |\nabla u|^2 \chi_{G^+} g \, dm]$$

where $\delta > 0$ is at our disposal.

To recapitulate, we have the estimate

$$\int |\nabla u|^2 \chi_{G^+} g \, dm = \int \nabla(u^2 \chi_{G^+}) g \, dm + \text{error}$$

$$\leq c \int_{\{(N(u) \leq \lambda\}} N^2(u) \, dm_1 + c \, \lambda^2 \, m(N(u) > \lambda) + \text{error}$$

$$\leq c \int_{\{N(u) \leq \lambda\}} |N(u)|^2 \, dm_1 + c \, \lambda^2 \, m_1(N(u) > \lambda) + \delta \int |\nabla u|^2 \chi_{G^+} g \, dm.$$

If δ is small, we may subtract the last term from both sides of the inequality. Now recall that

$$\int A^2(u) \, dm_1 \leq c \int |\nabla u|^2 \tilde{\chi}_{G^+} \, g \, dm \text{ (Fubini)}$$

$$\leq c \int_{\{N(u) \leq \lambda\}} N^2(u) \, dm_1 + c \, \lambda^2 \, m_1(N(u) > \lambda),$$

which is the desired estimate.

All of this work in the bidisque; that is essentially the observation of M.P. and P. Malliavin [6]. Of course, the calculations are much more complicated, and the estimation of the error term is more delicate. The interested reader may find the details in [4].

REFERENCES

[1]. Burkholder, D.L., Gundy, R.F., Extrapolation and interpolation of quasi-linear operators on martingales, Acta Math. 124 (1970) 249-304.

[2]. Burkholder, D.L., Gundy, R.F., and Silverstein, M.L. A maximal function characterization of the Class H^p, Trans. Amer. Math. Soc. 157 (1971), 137-153.

[3]. Fefferman, C., Stein, E.M., H^p spaces of several variables, Acta Math. 129 (1972) 137-193.

[4]. Gundy, R.F. Inégalitiés pour martingales à un et deux indices: L'espace H^p. L'Ecole d'Eté de St. Flour, 1978, Lecture Notes, Springer-Verlag. To appear.

[5]. Gundy, R.F., Stein, E.M. H^p theory for the polydisc, Proc.Nat. Acad. Sciences, USA, to appear.

[6]. M.P. and P. Malliavin, Intégrales de Lusin-Calderón pour les fonctions biharmoniques, Bull. Sci. Math. 101 (1977), 357-384.

HARMONIC ANALYSIS BASED ON CROSSED PRODUCT ALGEBRAS AND MOTION GROUPS

David Gurarie
Institute of Mathematics
The Hebrew University of Jerusalem

Introduction

There are three properties of commutative Banach algebras which play an important part in Harmonic Analysis. These are symmetry, the Wiener property and regularity.

Let us recall that a Banach-$*$-algebra $\underline{A}$ (not necessarily commutative) is called symmetric if the spectrum of every element $a^* \cdot a$ $(a \in \underline{A})$ is non-negative. An equivalent condition: the spectrum of every hermitian element $(h = h^*)$ is real.

A commutative Banach algebra $\underline{A}$ is called regular if the Gelfand transform $f \to \bar{f}(x)$ takes $\underline{A}$ into a regular class of functions on the maximal ideal space X of $\underline{A}$.

A Banach algebra $\underline{A}$ has the Wiener property (is Wiener) if every closed ideal $I \subset \underline{A}$ is contained in a maximal closed ideal.

The well-known example of an algebra satisfying all three properties is the group algebra $L^1(G)$ on a locally compact abelian group G or, more generally, weighted (Beurling) group algebra $L^1(G)$, with symmetric weight ρ, $\rho(g) = \rho(g^{-1})$ $(g \in G)$, whose growth is not "too" fast. As proved in [2] a sufficient and in certain sense a necessary condition is the non-quasi-analycity of ρ

$$\sum_{-\infty}^{\infty} \ln \rho(g^n)/(1+n^2) < \infty \quad \forall g \in G \tag{1}$$

The Wiener property enables us to define an important notion of the spectrum (or hull) $h(I)$ of an ideal $I \subset \underline{A}$ and to pose the problem of <u>spectral synthesis</u>. By definition $h(I)$ consists of all maximal ideals $M_x (x \in \hat{\underline{A}})$ which contain I.

The problem of spectral synthesis can be posed as follows

I) To describe the ideals of $\underline{A}$ with the "most simple" spectrum, i.e. one-pointed. They are called <u>primary</u> ideals.

II) To characterize the ideals of $\underline{A}$, which coincide with the intersection of all primary (maximal) ideals, which contain them, so-called <u>synthesizable ideals</u>.

Both these problems are rather complicated and most of the known results concern special cases of groups and algebras. Two general results should be mentioned. They are <u>Theorem of Ditkin</u>, which claims the absence of non trivial primary ideals in the group algebra $L^1(G)$ on a locally compact abelian group G, and Theorem of <u>Malliavin</u> which states the existence of non-synthesizable ideals in $L^1(G)$ on every locally compact non-compact abelian group G (for both of them see [8], Ch. 10).

Proceeding to the non commutative case no one of the above three basic properties have to stay true in general. The known counter-example is the group algebra $L^1(G)$ on a semisimple non compact Lie group, which is neither symmetric, nor Wiener, nor regular (see [10]).

Recently in a number of papers these properties (especially symmetry and Wiener) were investigated for different classes of groups. The positive answer was obtained for all concerned locally compact groups of polynomial growth [12], discrete solvable groups [12], certain classes of solvable Lie groups [11], motion groups [5], and some others.

Motion groups, which naturally generalize both commutative and compact groups are well known in representation theory by a number of "good" properties. One defines them as a semidirect product of a locally compact abelian group A by a compact group U acting on A by automorphisms. Therefore they prove to be "natural" candidates for study of the above harmonic analysis problems.

In the present paper we deal with two of them. The first one concerns extending the three basic properties to motion groups, or more generally to crossed product algebras. We shall prove Theorem 1: if a commutative Banach-*- algebra $\underline{A}$ is symmetric, Wiener and regular then the crossed product (or generalized twisted L^1 algebra) $\underline{L} = L(U,\underline{A})$ of $\underline{A}$ by a compact group U acting on $\underline{A}$ satisfies the same properties (for definition of these properties in the non-commutative case see §1).

As a consequence we get that these properties hold for any Beurling algebra $L^1_\rho(G)$ on a motion group G with a non-quasianalytic weight ρ.

The second problem we deal with concerns primary ideals of crossed products and group algebras of motion groups. As in the commutative case the Wiener property enables us to define the spectrum (hull) of an ideal and to ask: what are the primary ideals of group algebra $L^1(G)$ or, more generally, of crossed product $\underline{L} = L(U,\underline{A})$?

We state a sufficient condition on a pair $(\underline{A},U)$, which implies the Ditkin property i.e. the absence of nontrivial primary ideals in the crossed product $\underline{L} = L(U,\underline{A})$ (Theorem 3). Then we apply this condition in some special cases. In particularly, we get the absence of nontrivial primary ideals for group algebras $L^1_\rho(G)$ on motion groups over non archimedean fields (Corollary 4) and the Euclidean motion group $R^2 \lambda SO(2)$ (Corollary 5).

In contrast to this the primary ideals do exist for Euclidean motion groups $G = R^n \lambda SO(n)$ with $n \geq 3$ (Theorem 4). The dual space of such a group G breaks down into the union of two disjoint subsets: $\hat{G} = \hat{G}_+ \cup \hat{G}_o$, the "smooth part" $\hat{G}_+$ being a collection of semiaxes $\{R^+_\tau\}$. For any point $\lambda \in \hat{G}_+$ there exist a decreasing chain of $[\frac{n+1}{2}]$ primary ideals of the algebra $L^1(G)$, while at the points $\pi \in \hat{G}_o$ there are no nontrivial primary ideals. Moreover, their properties are similar to the properties of primary ideals in the commutative one-dimensional case cf. [16]). For instance, the k-th primary ideal in the chain coincides with the k-th power of the first (a maximal) one.

It is also described by the vanishing at the point $\lambda \in \hat{G}_+$ of the Fourier transforms $\hat{f}(\lambda)$ along with a certain number of their derivatives.

§1. Symmetry, regularity, the Wiener property

Let $\underline{A}$ be a Banach algebra and U be a locally compact group, acting continuously on $\underline{A}$ by automorphisms

$$a \to a^u (a \in \underline{A},\ u \in U)$$

Following |9, 15|, we define the crossed product (or generalized twisted L^1-algebra) $\underline{L} = \underline{L}(U,\underline{A})$ to be a B-algebra of all $\underline{A}$-valued measurable function on U with the norm

$$||f|| = \int_U |f(u)|du, \quad (f = f(u) \in \underline{L}) \tag{2}$$

and multiplication

$$(f_1 * f_2)(u) = \int_U f_1(v) \cdot [f_2(v^{-1}u)]^{v-1} dz. \tag{3}$$

If $\underline{A}$ is a $*$-algebra, then the correspondence

$$f \to f^* = [f(u^{-1})^*]^{u-1}$$

defines involution on $\underline{L}$.

The covariance representation of the pair $(\underline{A},U)$ see[9,15]) is by definition a pair of representations (L,S) of $\underline{A}$ and U correspondingly on the same Banach space E satisfying the relation

$$L_{a^u} = S_{u^{-1}} \cdot L_a \cdot S_u \qquad (a \in \underline{A},\ u \in U). \tag{4}$$

Any such pair defines a representation of the crossed product $\underline{L}$ in space E

$$T_f = \int_U L_{f(u)} \cdot S_u du, \quad (f = f(u) \in \underline{L}) \tag{5}$$

Conversely any nondegenerate representation of the crossed product $\underline{L}$ can be obtained in this way ([9]).

The natural examples of crossed products are group algebras of semidirect products of locally compact groups. Let $G = A \lambda U$, where U acts on A by automorphisms $u: a \to a^u$ $(a \in A, u \in U)$. We put $\underline{A} = L^1(A)$. The group U acts on $\underline{A}$ by changing the variable

$$f \to f^u(a) = f(a^{u-1}) \quad (a \in A, \quad u \in U). \tag{6}$$

It is easily seen that the group algebra $L^1(G)$ is isomorphic to the product $\underline{L} = L(U, L^1(A))$. In certain cases a Beurling group algebra $L^1_\rho(G)$ can also be realized in this way. For instance, if U is compact, any weight ρ on G is equivalent to the U-biinvariant one: $\rho(ugv) = \rho(g)$, $\forall g \in G$, $u,v \in U$. So ρ is actually defined by its restriction ρ' on A. Then $L^1_\rho(G) \simeq \underline{L}(U, L^1_{\rho'}(A))$.

From now on we shall restrict ourselves to the case when A is commutative and U is compact. The corresponding groups are semidirect products of a LCA group A on a compact group U, i.e. <u>motion groups</u>.

Unitary irreducible representation of motion groups are well-known and one can easily derive their description from Mackey's theory of group extensions [13]. In the paper [15] a similar theory was developed for crossed product C*-algebras and their *-representations.

We refer to [13,15] for exact definitions and results, and here we just notice that in our special case, the <u>unitary dual space</u> $\hat{\underline{L}}_*$ of a crossed product $\underline{L} = \underline{L}(U,A)$ can be identified with the set of all pairs $\{(\omega,\tau)\}$, ω being an orbit of U in the dual (Gelfand) space $\hat{\underline{A}}$ of $\underline{A}$, τ being an irreducible (unitary) representation of a stability subgroup U_x of some point $x \in \omega$ $(\tau \in \hat{U}_x)$. The irreducible representation $T^{(\omega,\tau)}$ is induced by a finite dimensional representation $\chi \otimes \tau$ of a subalgebra $\underline{L}_\chi = \underline{L}(U_\chi, A)$ (see [15]).

Proceeding to the nonunitary case one needs to modify some basic notions such as irreducibility, equivalence, dual space and so on. The class of representations we deal with are Banach representations, i.e. representations of groups (algebras) by bounded operators on a Banach space E. By irreducibility we shall always mean a complete one, saying that the associative hull of a representation is weakly dense in the algebra of all bounded operators on E.

Given an algebra $\underline{L}$ and a completely irreducible representation T of $\underline{L}$ we call its kernel $N_T = \{f \in \underline{L} \mid T_f = 0\}$ a primitive ideal of $\underline{L}$. The dual space of $\underline{L}$ is by definition $\hat{\underline{L}}$ = Prim $\underline{L}$ (cf. [3,4]).

We also introduce two other dual spaces: Max $\underline{L}$ - the set of all maximal two-sided ideals of $\underline{L}$, and $\mathrm{Prim}_*\underline{L}$ the set of all *-primitive ideals. The equality of kernels defines an equivalence relation on the set of all irreducible representations of $\underline{L}$ which is called in [4] a functional equivalence. In some cases it coincides with the usual equivalence, for instance, if $\dim T < \infty$ or T is a unitary or *-representation in Hilbert space. For an important class of algebras, which contain "enough" small idempotents (see [3]), in particular for locally compact groups which contain a massive compact subgroup, it coincides with the weaker yet rather natural Naimark equivalence [3].

Now, given a locally compact group G and a weight ρ on G let us define $\hat{G}$ as a dual space of the group algebra $L^1_\rho(G)$ (cf. [3]). In a similar way we define the dual spaces $\mathrm{Max}_\rho G$ and $\mathrm{Prim}_* G$. It is easily seen that $\hat{G}_\rho$ consists of the equivalence classes of irreducible representations T of G, whose growth is less than ρ, $||T_g|| \leq \rho(g)$, $\forall g \in G$.

The dual space $\hat{\underline{L}}$ of an algebra $\underline{L}$ can be topologized in two different ways. The first one gives the well-known hull-kernel or Jacobson topology, while the second one introduced by J. Fell [4] defines a so-called functional topology, the one which naturally generalizes the Gelfand topology on the maximal ideal space of a commutative Banach algebra. We shall refer to them as J- and F-topologies.

In general, F-topology is stronger than the J-topology. They are known to be equivalent for regular commutative Banach algebras. As shown in [4] by J. Fell, this is also true for C*-algebras. Therefore we can give the following definition of regularity.

Definition 1. Algebra $\underline{L}$ is called regular if J- and F-topologies on its dual space $\underline{L}$ are equivalent.

Scch algebras retain some important properties of commutative regular algebras.

Definition 2. (cf.[10, 5]) Algebra $\underline{L}$ is called Wiener, if any two-sided ideal $I \subset \underline{L}$ is contained in a primitive ideal M_λ $(\lambda \in \hat{\underline{L}})$.

Theorem 1. Let $\underline{A}$ be a commutative symmetric regular and Wiener B-algebra, which contains a bounded approximate identity. If $\underline{L} = \underline{L}(U,A)$ is a crossed product of $\underline{A}$ by a compact group U, then $\underline{L}$ retains all basic properties of $\underline{A}$. Namely,

a) $\underline{L}$ is symmetric regular and Wiener.

b) $\mathrm{Prim}\,\underline{L} = \mathrm{Prim}_*\underline{L} = \mathrm{Max}\,\underline{L}$,

Algebra $\underline{L}$ will be shown in the proof of Theorem 1 to have "enough" small idempotents in the sense of [3]. So, as noted above, a functional equivalence coincides with the Naimark one. Thus, we get

Corollary 1. Each irreducible representation T of $\underline{L}$ is Naimark equivalent to some of its irreducible *-representations in Hilbert space.

Applying the above results and those of [2] to motion groups we get.

Theorem 2. If $G = A_\lambda U$ is a motion group and a weight ρ on G is nonquasianalytic, then

a) The group algebra $L^1_\rho(G)$ is symmetric regular and Wiener

b) $\hat{G} = \mathrm{Max}_\rho G = \mathrm{Prim}_* G = \hat{G}_*$ (since G is a type I group)

In particular,

c) each Banach irreducible representation T of G of a non-quasianalytic growth (i.e. $\rho(g) = ||T_g||$ satisfies (1)) is Naimark equivalent to a unitary irreducible representation of G.

This theorem generalizes the results of [2] and [5].

Remark. As we noted in the commutative case the nonquasianalyticity of ρ is in a certain sense a necessary condition for the three basic properties to hold for $L^1_\rho(G)$ (see [17]). It does so (in a "similar sense") for motion groups. Namely, let $G = R^n \lambda SO(n)$ be a Euclidean motion group and let a weight ρ on G satisfy some "regular-growth" condition at ∞ (see [17]). Then if ρ is quasianalytic.

$$\sum_{-\infty}^{\infty} \rho(g^n)/(1+n^2) = \infty, \quad \text{for some } g \in R^n,$$

$L^1(G)$ can be shown to violate all three basic properties.

Let us outline the proof of Theorem 1. The principal point concerning $\underline{L} = \underline{L}(U,A)$ that we make use of is the existence in $\underline{L}$ of "enough small idempotents" (see [3]). An idempotent $e \in \underline{L}$ is called small if $\mathrm{rk}\, T_e \leq m = m(e)$ for any irreducible $T \in \hat{\underline{L}}$. An algebra $\underline{L}$ is said to have enough smal idempotent if they generate a dense two sided ideal $\underline{A}_0 \subset \underline{A}$. We shall prove that in our case the matrix elements $\{\pi_{ii}\}_{i=1}^{d(\pi)}$ of irreducible representations $\pi \in \hat{U}$ are small idempotents of $\underline{L}$*)

Let us correspond to an imdempotent $e \in \underline{L}^*$ a subalgebra of "spherical functions" $L(e) = e * \underline{L} * e$. There exists a close connection between irreducible representations of $\underline{L}$ and $\underline{L}(e)$ as well as between their two-sided and one-sided (maximal and primitive) ideals

*) Actually they don't belong to $\underline{L}$ provided $\underline{A}$ has no identity, but since $\underline{L}$ has a "natural" structure of (left and right) $L^1(U)$-bimodule (see [9]) we can "multiply" elements $f \in \underline{L}$ by $\psi \in L^1(U)$ ($\psi * f$ and $f * \psi \in \underline{L}$).

(see [6,3]). In particular the dual space $\underline{L}(e)$ is naturally identified with an open (in J-topology) subset $Q = \{T \in \hat{\underline{L}} | T_e \neq 0\}$ of $\hat{\underline{L}}$. When e $\underline{L}$ runs over a "sufficient system of idempotents" we get an open covering of $\hat{\underline{L}}$ by subsets $\{\hat{\underline{L}}(e)\}_{e\in\Delta}$.

This yields the reduction of such properties as Wiener, regularity and the equality of dual spaces (Theorem 1;b) from the whole $\underline{L}$ to its subalgebras $\{L(\pi), \pi\in\hat{U}\}$.

For the latters we get the following realization. Let $\underline{B} = \underline{A} \otimes M_n$ be a product of $\underline{A}$ by $n \times n$ - complex matrix algebra $M_n (n = d(\pi))$. The group U acts on $\underline{B}$ by tensor product of its action on $\underline{A}$ and a conjugation $u: s \to \pi(u)^{-1} \cdot s \cdot \pi(u)$ $(s \in M_n, u \in U)$ on M_n.

<u>Lemma 1</u>. A subalgebra $\underline{L}(\pi) \subset \underline{L}$ is isomorphic to the algebra of all U-invariants of $\underline{B}$,

$$\underline{L}(\pi) = \{F = (f_{ij})_{ij=1}^{n} \ (f_{ij} \in \underline{B}) | \quad (f_{ij}^{u}) = \pi(u)^{-1} \cdot (f_{ij}) \cdot \pi(u)\} .$$

Using a matrix-valued Fourier-Gelfand transform $\underline{F}: \underline{B} \to C(X) \otimes M_n$ $(X = \hat{\underline{A}})$, we represent $\underline{L}(\pi)$ by <u>U-invariant</u> vector-valued functions on X,

$$\hat{F}(x^u) = \pi(u)^{-1} \cdot \hat{F}(x) \cdot \pi(u) \qquad (F \in L(\pi) \subset \underline{B}; \quad u \in U) . \tag{7}$$

In particular the values of $\hat{F}(x)$ on each orbit $\omega \subset X$ are uniquely defined by its value at a single point $x_0 \in X$. For any $x \in X$ the matrices $\{\hat{F}(x) | F \in \underline{L}(\pi)\}$ belong to the commutant $\underline{R}'_x$ of the algebra $\underline{R}_x$ generated by the restriction of π on a stability subgroup U_x of x.

<u>Lemma 2</u>. For any $x \in X$ the image of $\underline{L}(\pi)$ at x is equal to R'_x.

As it is well known, both algebras $\underline{R}_x$ and $\underline{R}'_x$ are semisimple, so they split into the direct sum of full matrix algebras.

$$\underline{R}_x = \bigoplus_\tau M_{d(\tau)} \otimes I_{k(\tau)} \; ; \; \underline{R}'_x = \bigoplus_\tau I_{d(\tau)} \otimes M_{k(\tau)} . \qquad (8)$$

Here $\tau \in \hat{U}_x$ runs over all primary components of a restriction $\pi \mid U_x \simeq \bigoplus_\tau \tau \otimes k(\tau)$, $d(\tau)$ and $k(\tau)$ standing for its degree and multiplicity in $\pi \mid U_x$.

Thus, we can attach to any pair (ω,τ) $(\omega \subset X, \tau \in \hat{U}_x$ for some $x \in \omega)$ a class of equivalent finite dimensional representations $\lambda_{\omega,\tau}$

$$\lambda_{\omega,\tau}(F) = \hat{F}_\tau(X), \quad (x \in \omega)$$

$\hat{F}_\tau(x)$ being a τ-component of a decomposition $\hat{F}(x) = \bigoplus_\tau I_{d(\tau)} \otimes \hat{F}_\tau(x)$. The kernel $N_{\omega,\tau}$ of $\lambda_{\omega,\tau}$ is a two-sided maximal and primitive ideal of $\underline{L}(\pi)$.

<u>Lemma 3</u>. Each proper closed two-sided ideal of $\underline{L}(\pi)$ is contained in some $N_{\omega,\tau}$.

To prove Lemma 3 we consider a subalgebra $\underline{Z}$ of U-invariants of $\underline{A}$, $\underline{Z} = \{a \in \underline{A} \mid a^u = a, \; \forall u \in U\}$. It is identified with a central subalgebra of $\underline{L}(\pi)$ which consists of all U-invariant scalar matrix-functions on X.

The subalgebra $\underline{Z}$ retains all basic properties of $\underline{A}$. In particular it is regular, Wiener and contains an approximate identity of the whole $\underline{A}$. Using these properties one can show that if a two-sided ideal $J \subset \underline{L}(\pi)$ is contained in no $N_{\omega,\tau}$ then $J \supset \underline{Z}$ and hence $J = \underline{L}(\pi)$

Lemma 3 yields a description of the dual space $\hat{\underline{L}}(\pi)$ as the set of all pairs $\{(\omega,\tau)\}$. It also yields the Wiener property and the equality of dual spaces: Prim $\underline{L}(\pi)$ = Max $\underline{L}(\pi)$ = $\text{Prim}_*\underline{L}(\pi)$. As a consequence of [6,3] we get these properties for the whole $\underline{L}$.

Let us also note that for any $\lambda \in \hat{\underline{L}}(\pi)$ its degree $d(\lambda) \leq d(\pi)$, i.e. $\underline{L}(\pi)$ is a Banach *-algebra with the dual space of <u>bounded degree</u> see [4]), in particular all idempotents $\{\pi_{ii}\}_i$ $(\pi \in \hat{U})$ are small.

We prove the regularity of $\underline{L}(\pi)$ using the results of [4] concerning this class of algebras and the following:

Lemma 4. The correspondence $p : (\omega,\tau) \to \omega$ from $\hat{\underline{L}}(\pi)$ onto the space of orbits $\Omega = X/U$ is continuous in J-topology, and the inverse image $p^{-1}(\Delta)$ of any compact subset $\Delta \subset \Omega$ is compact in F-topology.

The regularity of all subalgebras $\underline{L}(\pi)$ yields, via the correspondence of dual spaces $\underline{L}(\pi)$ and $\hat{\underline{L}}$ (see [3]), the regularity of $\underline{L}$.

The symmetry of $\underline{L} = \underline{L}(U,\underline{A})$ in a more general context was recently proved by H. Leptin and D. Poguntke [11]. They proved that a crossed product of any (not necessarily commutative) symmetric algebra $\underline{A}$ by a compact group U is also symmetric.

The Theorem is thus proved.

We conjecture that Theorem 1 should hold for crossed products $\underline{L}(U,\underline{A})$ with "appropriate" noncommutative algebras $\underline{A}$ as well.

§2. The primary ideals of crossed products and motion groups

The Wiener property as in the commutative case enables us to introduce a notion of the spectrum (or hull) $h(J)$ of a two-sided ideal $J \subset \underline{L}$. By definition $h(J)$ consists of all maximal (primitive) two-sided ideals $\underline{M}_\lambda$ $(\lambda \in \hat{\underline{L}})$ which contain J.

From spectral viewpoint the most simple ideals are those with one-pointed spectrum. They are called primary ideals. One asks to describe primary ideals of a crossed product $\underline{L} = L(U,A)$. In general their structure seems to be rather complicated, so we shall restrict ourselves to special cases of algebras and groups.

Let us call algebra $\underline{L}$ Ditkin if it has no nontrivial primary ideals (the trivial ones are, of course, maximal).

We recall that a closed subset Δ of the Gelfand space X of a commutative B-algebra $\underline{A}$ is called synthesizable, if the smallest ideal $k_0(\Delta)$ with spectrum Δ is equal to the biggest one $k(\Delta)$.

Theorem 3. If a B-algebra $\underline{A}$ is symmetric regular Wiener and each orbit $\omega \subset X = \hat{\underline{A}}$ is synthesizable, then the crossed product $\underline{L} = \underline{L}(U,A)$ is Ditkin.

The proof of Theorem 3 involves considering the "generalized spherical" subalgebras of $\underline{L}$ (so called block-algebras [3,4]). For a finite subset $\Sigma = \{\pi_1, \ldots \pi_m\} \subset \hat{U}$ we define an idempotent $\chi_\Sigma = \sum_1^m d(\pi_i)\, tr\pi_i$ and set $\underline{L}(\Sigma) = \chi_\Sigma * L * \chi_\Sigma$. These algebras retain all of the basic properties of $\underline{A}$ and $\underline{L}(\pi)$. In particular, they are symmetric, regular, Wiener and each has a dual space of bounded degree,

$$d(\lambda) \leq \sum_{i=1}^{m} d(\pi_i), \qquad \lambda \in \hat{\underline{L}}(\Sigma).$$

Their dual spaces $\hat{\underline{L}}(\Sigma)$ can be identified with open subsets of $\hat{\underline{L}}$. Moreover, the block-algebras $\underline{L}(\Sigma)$ "approximate" $\underline{L}$, when Σ runs over all finite subsets of $\hat{U}$, $\underline{L} = \overline{\{\cup \underline{L}(\Sigma) \mid \Sigma \subset \hat{U}\}}$.

Correspondingly, any two-sided ideal $J \subset \underline{L}$ is "approximated" by intersections $J(\Sigma) = J \cap \underline{L}(\Sigma) = \chi_\Sigma * J * \chi_\Sigma$.

So, we have only to establish the Ditkin property for all subalgebras $L(\Sigma)$. Once again we consider a subalgebra of U-invariants $\underline{Z} \subset \underline{A}$, identified with a central subalgebra of $\underline{L}(\Sigma)$, and use the following important property of regular B-algebras $\underline{B}$ of bounded degree: for any T-closed subset* $\Delta \subset \hat{\underline{B}}$ (i.e. a subset which coincides with the intersection of all its closed neighbourhoods), there exist a smallest ideal $k_0(\Delta)$ of $\underline{B}$ with spectrum Δ. This property is well known to characterize the class of commutative regular $\underline{B}$-algebras, "T-closed" being the usual "closed".

Let us give some applications of Theorem 3.

Corollary 2. If $\underline{A} = C(X)$ is the B-algebra of all continuous bounded functions on X (or its subalgebra of the functions going to 0

---

*Let us notice that the dual space of noncommutative B-algebras in general is non-Hausdorff. But in our case $\hat{L}(\pi)$ (and $\hat{L}$ itself) satisfies the T_1-axiom.

at ∞), then the crossed product $\underline{L} = L(U,\underline{A})$ with any compact group U is Ditkin.

Corollary 3. If $G = A \lambda U$ is a motion group with finite orbits $\omega \subset \hat{A}$ (in particular, if U is finite) then the group algebra $L^1(G)$ is Ditkin.

The synthetizibility of finite subsets $\omega \subset \hat{A}$ follows from Ditkin's theorem ([8], ch. 10).

Another class of groups to be considered are motion groups over non archimedian fields K, $G = K^n \lambda SL(n,Q)$, where Q is a ring of integers of K.

Corollary 3. If G is a non archimedian motion group, then a Beurling group algebra $L^1_\rho(G)$ with any weight ρ is Ditkin.

Indeed, each orbit $\omega \subset \hat{A} = K^n$ is either a sphere of radius $r = p^{mk}$ $(k = 0; \pm 1, \ldots)$ or the point $\{0\}$. Obviously, each non zero ω is both an open and closed subset of $\hat{A}$, so $k_0(\omega) = k(\omega)$. As for the point $\{0\}$ one can use the following version of Ditkin's theorem: if a LCA group A coincides with the union of its compact subgroups, then a group algebra $L^1_\rho(A)$ with any weight ρ is Ditkin.

Corollary 5. If $G = R^2 \lambda SO(2)$ is a Euclidean motion group, then the group algebra $L^1(G)$ is Ditkin.

The synthetizability of circles $\omega_r \subset R^2$ was proved by C. Herz [7].

Let us notice that we cannot apply Theorem 3 to Euclidean motion groups $G = R^n \lambda SO(n)$ with $n \geq 3$, since, as well known (see [16]) a sphere S_r $(r > 0)$ in R^n $(n \geq 3)$ is not synthetizable. It turns out that the group algebra $L^1(G)$ of a Euclidean motion group does have primary ideals and their structure is similar to the structure of primary ideals of a subalgebra $\underline{Z}$ of rotationally invariant functions $f \in L^1(R^n)$ (cf. [16]). Proceeding to their study let us recall a description of the dual space $\hat{G}$ of Euclidean motion groups and some of its properties.

Each orbit $\omega \subset R^n$ is either a sphere ω_r of radius $r > 0$ or the point $\{0\}$. A stability subgroup of any point $x \in \omega_r$ $(r > 0)$ is isomorphic to $SO(n-1)$, while the stability subgroup of $\{0\}$ is $SO(n)$ itself. So the dual space $\hat{G}$ breaks down into the union of two disjoint subsets $\hat{G} = \hat{G}_+ \cup \hat{G}_0$, where

$$\hat{G}_+ = \{\lambda = (r,\tau) \quad r \in R^+, \quad \tau \in \hat{SO}(n-1)\} \cong R^+ \times \hat{SO}(n-1) = \bigcup_{\tau \in \hat{SO}(n-1)} R^+$$

is a union of disjoint open semiaxes and

$$\hat{G}_0 = \{\lambda = \pi \mid \pi \in \hat{SO}(n) \cong \hat{SO}(n) \quad \text{is discrete}$$

All irreducible representations T^μ $(\mu = (r,\tau) \in R^+_\tau)$ act on the same space $E^\tau = L^2(\omega) \otimes E(\tau)$, which is the space of an induced representation Ind (τ). The space E^τ decomposes into the direct sum of its U-primary components

$$E^\tau = \oplus_{\pi \in \hat{U}} E^\tau(\pi), \tag{9}$$

the restriction $U \mid E^\tau(\pi)$ being a multiple of an irreducible $\pi \in \hat{U}$. Actually one can show (see [18], ch. 18), that for any $\pi \in \hat{U}$ its multiplicity in $T^\mu \mid U$ is 1 or 0. We denote by $E^\tau_0 \subset E^\tau$ a dense subspace, which consists of all finite linear combinations of vectors $\{\xi \in E^\tau(\pi) \mid \pi \in \hat{U}\}$.

<u>Theorem 4</u>. Let $G = R^n \lambda SO(n)$ be a Euclidean motion group. Then

a) The group algebra $L^1(G)$ has no nontrivial primary ideals at any point $\{\pi\} \in \hat{G}_0$.

b) For any point $\lambda \in \hat{G}_+$ there exists a decreasing chain of $[\frac{n+1}{2}]$ primary ideals

$\underline{M}_\lambda = J_0(\lambda) \supset J_1(\lambda) \supset \ldots \supset J_{[\frac{n-1}{2}]}(\lambda)$. Moreover,

c) $J_k(\lambda) = \underline{M}_\lambda^{k+1}$ (the k+1)-th power of a maximal ideal).

d) When vectors ξ,ζ run over a subspace E_0^τ the matrix elements $\{\hat{f}_{\xi,\zeta}(\lambda) = \langle T_f^\lambda \xi,\zeta\rangle;\ f \in L^1(G)\}$ are differentiable $[\frac{n-1}{2}]$ times in $\lambda=(r,\tau)$ and

$$J_k(\lambda) = \{f \in L^1(G) \mid (\frac{d}{dr})^j \hat{f}_{\xi,\zeta}(\lambda) = 0;\ \forall\ \xi,\zeta \in E_0^\tau \text{ and } j = 0,1,\ldots,k\}.$$

We prove the Theorem by reducing it from the whole algebra $\underline{L} = L^1(G)$ to its "spherical" subalgebras $\{\underline{L}(\pi) \mid \pi \in \hat{U}\}$.

Namely, we prove the following

<u>Lemma 5</u>. For any point $\lambda \in \hat{\underline{L}}(\pi) \subset \hat{G}$ there exists a one-to-one correspondence between primary ideals of $\underline{L}(\pi)$ and $\underline{L}$ with the spectrum

Then we notice that $\underline{L}(\pi)$ is a commutative B-algebra, whose dual space $\hat{\underline{L}}(\pi)$ is identified with the union of a finite number of semiaxes $\{R_\tau^+\}_\tau$,

$$\hat{\underline{L}}(\pi) = \{\pi\} \cup \{R_\tau^+ \mid \forall\, \tau : \pi|SO(n-1) \supset \tau\}$$

The Fourier-Gelfand transform of $\underline{L}(\pi)$ is expressed by <u>matrix elements</u> $\{\langle T_g^\mu \xi,\zeta\rangle \mid \xi,\zeta \in E_0^\tau\}$ of irreducible representations T^μ $(\mu \in \hat{\underline{L}}(\pi))$. These elements were shown in [14] to develop in finite linear combinations of Bessel functions with integer (or semiinteger) indices, depending, of course, on $(\xi,\zeta,\tau$ and $\pi)$.

So to establish a "sufficient" smoothness of functions $\{\hat{f}(\mu) \mid f \in \underline{L}(\pi)\,(\mu \in R_\tau^+)$ on the Gelfand space $\hat{\underline{L}}(\pi) = \bigcup_\tau R_\tau^+$ and to estimate the exact number of their continuous derivatives (which is just equal to the number of primary ideals of $\underline{L}(\pi)$ at the point $\lambda \in R_\tau^+$) we can use the expansions [14] along with the known results concerning the derivatives of Bessel functions and their asymptotic behaviour (see [1], ch. 7).

Thus, we obtain the structure of the primary ideals of $\{\underline{L}(\pi)\}$ and the properties c) and d) for them. By Lemma 5 we also obtain the structure of the primary ideals of the whole algebra $\underline{L} = L^1(G)$. To get the properties c) and d) for it we use the above arguments along with the results of [6,3] concerning the correspondence between two-sided ideals of algebra $\underline{L}$ and its subalgebras $\{\underline{L}(\pi)\}$ generated by small idempotents $\{\pi_{ii} | \pi \in \hat{U}\}$.

REFERENCES

[1]. H. Bateman, A. Erdelyi, Higher trancedental functions, 6.II McGraw-Hill, 1953.

[2]. Y. Domar, Harmonic analysis based on certain commutative Banach algebras, Acta Math. 96 (1956) 1-66.

[3]. J.M.G. Fell, Non unitary dual space of groups, Acta Math. 114 (1965), 267-310.

|4|. J.M.G. Fell, The dual space of Banach algebras, Trans.Amer. Math.Soc. 114 (1965), 227-250.

[5]. R. Gangolli, On the symmetry of L_1-algebras of locally compact motion groups and Wirner Tauberian theorem, J. Funct. Anal. 25 (1975), 224-252.

[6]. R. Godement, A theory of spherical functions, I, Trans. Amer. Math. SAc. 73 (1952).

[7]. C.S. Herz, Spectral synthesis for the circle, Ann.Math. 68 (1958), 709-712.

[8]. E. Hewitt, K.A. Ross, Abstract harmonic analysis, V. II, Springer Verlag, 1970.

[9]. H. Leptin, Verallgemeinerte L^1-Algebren und projektive Darstellungen lokal kompakter Gruppen, Inventiones Math., 3 (1967), 257-281; 4 (1967), 68-86.

[10]. H. Leptin, Ideal theory in group algebras of locally compact groups, Invent. Math. 31 (1976), 259-278.

[11]. H. Leptin, D. Poguntke, Symmetry and nonsymmetry for locally compact groups, Preprint.

[12]. J. Ludvig, A class of symmetric and a class of Wiener group algebras. Preprint.

[13]. G.W. Mackey, Inducted representations of locally compact groups. I. Ann. of Math. 55 (1952), 101-139.

[14]. L.V. Rosenblum, A.V. Rosenblum, On matrix elements of irreducible unitary representations of Euclidean motion group M(n). Russian Math. Survey (Uspechi) 29 (1974), N.4 (Russian).

[15]. M. Takesaki, Covariant representations of C -algebras and their locally compact groups, Acta Math. 119 (1967), 273-303.

[16]. N.T. Varopoulos, Spectral synthesis on spheres, Proc. Comb. Phil. Soc. (1966), 62, 379-387.

[17]. A. Vretblat, Spectral analysis in weighted L^1-spaces on R, Ark. f. Math. 11, 1973, 109-138.

[18]. D.P. Zelobenko, Compact Lie group and their representations, AMS, Providence 1973.

SUR LE TREIZIEME PROBLEME DE HILBERT, LE THEOREME DE SUPERPOSITION DE KOLMOGOROV ET LES SOMMES ALGEBRIQUES D'ARCS CROISSANTS

par J.P. Kahane

Il existe d'excellents exposés de synthèse sur le 13ème problème de Hilbert et la représentation des fonctions au moyen de "superpositions" (c'est-à-dire de compositions au sens $f \circ g$). Le plus récent est celui de Vitushkin [1], que nous alons suivre de très près. L'énoncé du problème et une mise au point par Lorentz se trouvent dans le volume consacré par l'American Mathematical Society aux problèmes de Hilbert [2].

Le sujet a certains rapports avec l'analyse de Fourier, en même temps qu'avec d'autres branches des mathématiques. C'est la raison pour en avoir donné un aperçu au colloque d'Héraclion.

Je me propose de présenter ici les questions suivantes, avec des résultats bien connus pour les premières et nouveaux pour les dernières.

1. existence de fonctions régulières de n variables non représentables comme superpositions de fonctions régulières de $n-1$ variables (théorèmes de Hilbert et de Vitushkin) (parties I, II)

2. représentation des fonctions continues de n variables par superpositions de fonctions d'une variable et d'additions (théorème de Kolmogorov) (partie III)

3. rôle des sommes algébriques de n arcs croissants génériques dans R^{2n+1} comme ensembles d'interpolation en analyse harmonique (partie IV)

4. rôle critique de la dimension $2n+1$ (partie V).

Les deux premières questions se trouvent exposées, avec plus ou moins de détails qu'ici, par Vitushkin dans [1]. La troisième a déjà été beaucoup étudiée ([3] à [10]); on donne ici des résultats nouveaux sur les ensembles de Helson et on signale le rôle des séries de Taylor uniformément convergentes, en application d'un théorème d'Alpár [11]. La quatrième est inspirée par le travail de R. Doss concernant le cas $n=2$ [12]; nous allons moins loin que Doss, mais nous avons un résultat générique intéressant, sur l'existence de points doubles dans une somme algébrique de n arcs croissants génériques dans R^m - je dois une partie de la démonstration à A. Fathi et F. Laudenbach - .

1. LE PROBLEME ET LE THEOREME DE HILBERT

Le 13ème problème de Hilbert a son origine dans la théorie des équations algébriques. Jusqu'au degré 6 inclus, les équations algébriques peuvent se résoudre en superposant des fonctions de deux variables. Au moyen de fonctions d'une variable et d'additions, la solution de l'equation générale du 7ème degré se ramène à celle de l'équation $X^7 + xX^3 + yX^2 + zX + 1 = 0$, c'est-à-dire à une fonction des trois variables x, y, z. Hilbert conjecture qu'on ne peut pas aller plus loin, c'est-à-dire qu'on ne peut pas obtenir cette solution en superposant des fonctions de deux variables. Il dit à ce sujet avoir démontré l'existence de fonctions analytiques de trois variables non représentables par superposition de fonctions de deux variables. Il est très probable qu'Hilbert n'avait en vue que des fonctions analytique de deux variables (voir théorème ci-dessous).

Cependant le 13ème problème de Hilbert - ou tout au moins l'une des versions de ce problème - s'énonce couramment ainsi: est-il vrai qu'existent des fonctions continues de trois variables x_1, x_2, x_3 $(0 \leq x_j \leq 1)$ qui ne puissent s'exprimer comme superpositions de fonctions continues de deux variables ? La réponse est alors négative (Arnold, Kolmogorov 1957) et nous y reviendrons.

Précisons ce qu'on entend par superposition d'ordre s de fonctions de p variables appartenant à une classe C, opérant sur $x_1, x_2, \ldots x_n$.

Les superpositions d'ordre 0 sont les fonctions $x_1, x_2, \dots x_n$. Si $U_1, U_2, \dots U_p$ sont des superpositions d'ordre $s-1$ et si $\varphi(t_1, t_2, \dots t_p)$ est une fonctions de la classe C, $\varphi(U_1, U_2, \dots U_p)$ est une superposition d'ordre s. La définition se transcrit immédiatement pour des germes de fonctions ou pour des séries formelles.

Par exemple, prenons pour C la classe des polynômes de deux variables, opérant sur x_1, x_2, x_3. Considérons les superpositions d'ordre s, modulo les monômes de degré strictement supérieur à m. Elles sont identifiables à des polynômes de degré m en x_1, x_2, x_3, et elles dépendent d'une nombre de paramètres que nous désignerons par $\gamma(m,s)$. Ainsi $\gamma(0,s) = 1$ et $\gamma(1,s) = 4$ pour tout $s \geq 1$. Le calcul de $\gamma(m,1)$ est un exercice combinatoire. Pour $k \geq 1$, il existe $3k$ monômes de degré k en x_1, x_2, x_3 où figurent seulement deux des variables; d'où

$$\gamma(m,1) = 1 + 3 + 3\times 2 + \dots + 3m = 1 + 3\,\frac{m(m+1)}{2}\,. \tag{1.1}$$

Pour $\gamma(m,s)$, contentons nous d'une majoration. Pensons à U_1 et U_2 comme à des superpositions arbitraires d'ordre $s-1$; modulo les termes de degré $>\mu$, l'une et l'autre dépendent de $\gamma(\mu, s-1)$ paramètres. Pensons à φ comme à un polynôme arbitraire de deux variables, et désignons par φ_k la somme des monômes de degrè k dans φ:

$$\begin{aligned}\varphi(X,Y) &= a + (bX + cY) + (dX^2 + eXY + fY^2) + \dots \\ &= \varphi_o(X, Y) + \varphi_1(X, Y) + \varphi_2(X, Y) + \dots\ .\end{aligned}$$

Ainsi $\varphi_k(X, Y)$ dépend de $k+1$ paramètres. Le nombre de paramètres dont dépend $\varphi\,(U_1, U_2)$ est majoré par

$$1+2+\dots+(m+1)+2\,\gamma(m,s-1)$$

On a donc

$$(1.2) \qquad \gamma(m,s) \leq \frac{(m+1)(m+2)}{2} + 2\,\gamma(m,s-1)$$

A partir de (1.1) et (1.2), il est immédiat de montrer que

$$(1.3) \qquad \gamma(m,s) \leq 2^{s-1}\left(\gamma(m,1) + \frac{(m+1)(m+2)}{2}\right) \leq 2^{s}(m+2)^{s}$$

Les polynômes de degré $\leq m$ en x_1, x_2, x_3 dépendent de d_m paramètres,

$$(1.4) \qquad d_m = \frac{(m+1)^3}{6} + \frac{(m+1)^2}{2} + \frac{m+1}{3}$$

Ils constituent donc une variété P_m de dimension d_m. Les sommes des monômes de degré m dans les superpositions d'ordre s de polynômes à deux variables opérant sur x_1, x_2, x_3 constituent une sous-variété $Q_{m,s}$ de P_m, de dimensions $\gamma(m,s)$. En composant (1.3) et (1.4), on voit que, pour m assez grand, on a

$$(1.5) \qquad d_m > \gamma(m, s) + 3.$$

On définit m_s comme le premier entier à partir duquel l'inégalité (1.5) est vraie. C'est la clé du théorème de Hilbert, que nous énonçons ainsi.

THEOREME. Quasi toute fonction entière $F(x_1, x_2, x_3)$ de trois variables a la propriété suivante: quel que soit $\xi = (\xi_1, \xi_2, \xi_3) \in \mathbb{R}^3$, le germe de F en ξ ne peut pas s'obtenir par superposition de séries formelles de deux variables.

"Quasi toute fonction entière" signifie: toute fonction entière appartenant à une intersection dénombrable d'ouverts denses dans l'espace E (métrique et complet) des fonctions entières de trois variables. On peut aussi dire "génériquement, toute fonction entière".

Le germe de F en ξ est la série de Taylor

$$S_\xi : \Sigma a_k(\xi) x^k \quad (k=(k_1,k_2,k_3),\ x=(x_1,x_2,x_3),$$
$$x^k = x_1^{k_1} x_2^{k_2} x_3^{k_3})$$

dont la somme est $F_\xi(x) = F(\xi + x)$. On posera

$$S_{\xi,m} : \sum_{|k| \leq m} a_k(\xi) x^k \qquad (|k| = k_1 + k_2 + k_3)$$

et, si l'on doit préciser, on écrira $S_{\xi,m}(F)$ au lieu de $S_{\xi,m}$. On montrera - ce qui est un peu plus fort que l'énonce du théorème - que, génériquement, pour tout ξ et tout s il existe un m tel que $S_{\xi,m}$ n'appartienne pas à $Q_{m,s}$.

Preuve du théorème. Pour tout entier $K > 0$, soit $Q_{m,s}(K)$ la partie de $Q_{m,s}$ constituée par les polynômes dont les coefficients sont majorés en module par K. Pour m, s, K donnés, soit $E(m,s,K)$ l'ensemble des fonctions entières F pour lesquelles

$$\forall \xi \quad |\xi| \leq K \Rightarrow S_{\xi,m} \notin Q_{m,s}(K).$$

$E(m,s,K)$ est un ouvert. En effet, en définissant sur P_m une métrique, la distance de $S_{\xi,m}$ au compact $Q_{m,s}(K)$ est une fonction continue de ξ et de F, et $E(m,s,K)$ est l'ensemble des F telles que cette distance ne s'annule pas sur le disque $|\xi| \leq K$.

$E(m,s,K)$ est dense si $m \geq m_s$. En effet, tout ouvert dans E contient, si ρ est assez petit, un translaté $F_o + P_m(\rho)$ de

$P_m(\rho)$ ($P_m(\rho)$ est l'ensemble des polynômes de degré $\leq m$ dont les coefficients sont de module $\leq \rho$). Soit $F = F_o + P$, $P \in P_m(\rho)$. La condition $S_{\xi,m} \in Q_{m,s}(K)$ s'écrit

$$S_{\xi,m}(P) \in Q_{m,s}(K) - S_{\xi,m}(F_o)$$

c'est-à-dire

$$P_\xi \in Q_{m,s}(K) - S_{\xi,m}(F_o)$$

c'est-à-dire que P appartient à une variété V_ξ de dimension $\gamma(m,s)$ dans P_m. La réunion des V_ξ est une variété de dimension $\gamma(m,s)+3$. Si $m \geq m_s$, la condition (1.5) a lieu et garantit l'existence d'un $P \in P_m(\rho)$ hors de $\bigcup_\xi V_\xi$. Alors $F_o + P \in E(m,s,K)$.

Soit $F \in \bigcap_{s,K} E(m_s,s,K)$ (intersection dénombrable d'ouverts denses). Pour tout s, on a $F \in \bigcap_K E(m_s,s,K)$ c'est-à-dire

$$\forall \xi \quad S_{\xi,m_s} \notin Q_{m_s,s} \qquad \text{CQFD.}$$

REMARQUES

1. On peut naturellement dans le théorème considérer au lieu de l'espace E de toutes les fonctions entières n'importe quel espace de fonctions entières, métrique et complet, tel que 1) les $a_k(\xi)$ soient fonctions continues de ξ et de F 2) tout ouvert contienne un translaté de $P_m(\rho)$ si ρ est assez petit. Par exemple, on peut imposer des conditions de croissance.

2. Soit $\alpha = \{\alpha_\varkappa\}$ $(\varkappa = 0,1,2,\ldots)$ une suite strictement positive tendant vers 0 plus vite que toute exponentielle ($\lim_{\varkappa\to\infty} \alpha_\varkappa A^\varkappa = 0$ pour tout $A > 0$). L'ensemble E_α des fonctions entières à trois variables

$$F(x) = \sum_k a_k x^k \quad (k = (k_1,k_2,k_3),\ x = (x_1,x_2,x_3),$$
$$x^k = x_1^{k_1} x_2^{k_2} x_3^{k_3})$$

telles que $|a_k| < \alpha_{|k|}$ ($|k| = k_1 + k_2 + k_3$) est un espace métrique complet. C'est aussi, de manière naturelle, un espace de probabilité, en considérant les a_k comme aléatoires, indépendants, et uniformément distribués sur $]-\alpha_{|k|}, \alpha_{|k|}[$. L'énoncé du théorème est alors valable en remplaçant "quasi toute F" par "presque toute F".

En effet, pour tout m, l'espace de probabilité E_α est le produit de deux espaces $P_m \cap E_\alpha$ et $R_m \cap E_\alpha$, R_m désignant l'ensemble des fonctions F telles que $a_k = 0$ pour $|k| \leq m$. Ecrivons $F = P + R$,

$$P(x) = \sum_{|k| \leq m} a_k x^k, \qquad R(x) = \sum_{|k| > m} a_k x^k.$$

Fixons s et $m = m_s$. Fixons R. L'ensemble des P tels qu'il existe $\xi \in \mathbb{R}^3$ pour lequel $S_\xi(P)$ appartient à $Q_{m,s} - S_\xi(R)$ est une variété de dimension $\gamma(m,s) + 3 < d_m$, donc est de probabilité nulle dans l'espace $R_m \cap E_\alpha$. Quand $s, m = m_s$ et R sont fixés, il est donc presque sûr que

$$\forall \xi \quad S_\xi(F) \notin Q_{m,s}.$$

Par le théorème de Fubini, cela reste vrai quand on fixe seulement s et $m = m_s$, et finalement il est presque sûr que

$$\forall s \ \forall \xi \quad S_\xi(F) \notin Q_{m_s,s}.$$

3. Le thèorème et les remarques 1 et 2 restent valables en remplaçant respectivement les fonctions de trois et deux variables par les fonctions de n et de $n-1$ variables $(n > 3)$. Il suffit dans la

preuve de modifier la définition de $\gamma(m,s)$ et de d_m et les évaluations (1.3) et (1.5).

4. Dans l'autre sens - en réduisant le nombre de dimensions - voici un résultat facile à obtenir. Quasi sûrement et presque sûrement (dans les conditions de la remarque 2) une fonction entière $F(x_1,x_2)$ a la propriété suivante: pour tout $\xi \in \mathbb{R}^2$ le germe de F en ξ ne peut pas s'obtenir comme superposition de séries formelles à une variable et d'additions.

5. Le théorème et les remarques ci-dessus restent valables en considérant $x_1,x_2,\ldots$ et $\xi_1,\xi_2,\ldots$ comme complexes.

6. Comme application du théorème, une fonction entière générique de 3 variables n'est au voisinage d'aucun point représentable comme superposition de fonctions C^∞ de deux variables.

7. Voici un théorème de Vitushkin (1964) (voir [1]) qui est beaucoup plus difficile. <u>Il existe une fonction analytique</u> $F(x_1,x_2)$ <u>qui n'est pas représentable sous la forme</u>

$$\sum_{i=1}^{N} p_i(x_1,x_2)\varphi_i(q_i(x_1,x_2))$$

<u>lorsque les</u> p_i et q_i <u>sont des fonctions fixées de deux variables de classe</u> C^1, <u>et les</u> φ_i <u>des fonctions continues arbitraires</u>.

Nous allons maintenent étudier un autre théorème de Vitushkin.

II. THEOREME DE VITUSHKIN (1954)

Soit $C_n^\alpha = C^\alpha(\mathbb{R}^n)$ l'ensemble des fonctions réelles de classe C^α de n variables réelles $(n = 2, 3, \ldots; \alpha > 0)$. De même, I^n désignant le cube unité $[0,1]^n$, et $\mathbb{T}^n$ le tore à n dimensions ($\mathbb{T} = \mathbb{R}/\mathbb{Z}$), on définit $C^\alpha(I^n)$ et $C^\alpha(\mathbb{T}^n)$. On va voir le rôle que joue le nombre $\frac{n}{\alpha}$ dans le problème de Hilbert, et dans la mesure de la "complexité" de l'espace C_n^α.

THEOREME. Supposons $\frac{n}{\alpha} > \frac{n'}{\alpha'}$, et $\alpha' \geq 1$. Alors quasi toute fonction dans $C^\alpha(I^n)$ (resp. $C^\alpha(\mathbb{T}^n)$) est non représentable par superpositions de fonctions de la classe $C_{n'}^{\alpha'}$ opérant sur $x_1, x_2, \ldots x_n$ (resp. sur $\cos t_1$, $\sin t_1$, $\cos t_2$, $\sin t_2, \ldots \cos t_n$, $\sin t_n$).

La meilleure démonstration de ce théorème de Vitushkin (1954) est fournie par la notion d' ε-entropie de Kolmogorov (1955). Nous allons la donner rapidement. Pour fixer les idées, et nous permettre l'usage des polynômes trigonométriques, nous allons travailler sur $\mathbb{T}^n$.

Soit K un compact dans l'espace de Banach $C(\mathbb{T}^n)$ des fonctions continues sur $\mathbb{T}^n$. Pour chaque $\varepsilon > 0$, $N(\varepsilon) = N(\varepsilon,K)$ désigne le nombre minimum de boules de rayon ε dont la réunion recouvre K. On appelle ε-entropie la fonction

$$H(\varepsilon) = H(\varepsilon,K) = \log N(\varepsilon) \qquad (\varepsilon < \varepsilon_0).$$

Choisissons une norme sur $C^\alpha(\mathbb{T}^n)$, et désignons par B_n^α la boule unité fermée de $C^\alpha(\mathbb{T}^n)$. C'est un compact dans $C(\mathbb{T}^n)$, et on peut montrer que

$$(2.1) \qquad H(\varepsilon, B_n^\alpha) \approx \left(\frac{1}{\varepsilon}\right)^{n/\alpha}$$

le signe $\approx$ signifiant que le rapport des deux membres est compris entre deux nombres strictement positifs (cf. |1|). Nous allons démontrer une version faible de (2.1) qui est suffisante pour le théorème, à savoir

$$(2.2) \qquad C_1\left(\frac{1}{\varepsilon}\right)^{n/\alpha} \leq H(\varepsilon, B_n^\alpha) \leq C_2\left(\frac{1}{\varepsilon}\right)^{n/\alpha}\log\frac{1}{\varepsilon} .$$

Pour la première inégalité (2.2) nous imitons |1|. On choisit un cube Q de côté $\frac{1}{2}$ dans T^n, et une fonction φ de classe C^∞, dont le support est intérieur à Q, et dont la norme $\|\varphi\|_\infty$ dans $C(T^n)$ égale 1. On partage Q en ν^n sous-cubes $Q_{\nu,j}$ de côté $\frac{1}{2\nu}$, et on désigne par $\varphi_{\nu,j}$ la transformée de φ par l'homothétie qui applique Q sur $Q_{\nu,j}$. En choisissant convenablement la norme dans $C^\alpha(T^n)$ (en prolongeant une norme naturelle de $C^\alpha(Q)$) on a pour tout ν et tout j

$$\|\varphi_{\nu,j}\|_\alpha = \nu^\alpha\|\varphi\|_\alpha ,$$

et on a additivité des normes pour des fonctions à supports disjoints dans Q. Donc, pour tout choix des signes $\pm$ on a

$$\|\sum_j \pm\varphi_{\nu,j}\|_\alpha = \nu^\alpha\|\varphi\|_\alpha .$$

Etant donné $\varepsilon > 0$, choissisons pour ν le plus grand entier tel que $\varepsilon\nu^\alpha\|\varphi\|_\alpha \leq 1$. Alors les 2^{ν^n} fonctions

$$\varepsilon\sum_j \pm\varphi_{\nu,j}$$

sont dans le boule B_n, et elles sont deux à deux à la distance $2\varepsilon\|\varphi\|_\infty = 2\varepsilon$ dans $C(T^n)$. Une boule de rayon ε dans $C(T^n)$ en contient au plus une, donc

$$N(\varepsilon, B_n^\alpha) \geq 2^{\nu^n}$$

et comme $\nu^\alpha \approx \frac{1}{\varepsilon}$ on a la première inégalité (2.2).

Pour la seconde inégalité (2.2) utilisons le théorème d'approximation de Jackson: pour chaque $f \in B_n^\alpha$ et chaque entier $\mu > 0$ il existe un polynôme trigonométrique T de degré $\leq \mu$ tel que

$$\|f - T\|_\infty \leq C\mu^{-\alpha} \qquad (C = C(\alpha)).$$

On peut supposer que les coefficients de Fourier de T sont bornés en module par 1, et leur nombre est inférieur à μ^n. Chaque T est donc approchable à moins de $\mu^n\delta$ près dans $\mathcal{C}(T^n)$ par un polynôme trigonométrique S de degré $\leq \mu$ dont les coefficients sont multiples de δ. Le nombre de ces polynômes trigonométriques S ne dépasse pas $(\frac{1}{\delta^2})^{\mu^n}$.

Etant donné $\varepsilon > 0$, choisissons pour μ le plus petit entier tel que $C\mu^{-\alpha} < \frac{\varepsilon}{2}$, puis δ de sorte que $\mu^n\delta = \frac{\varepsilon}{2}$. Toute f dans B_n^α est approchable à ε près par un polynôme trigonométrique S, donc

$$N(\varepsilon, B_n^\alpha) \leq (\frac{1}{\delta^2})^{\mu^n}$$

$$H(\varepsilon, B_n^\alpha) \leq \mu^n \log \frac{1}{\delta^2} \leq C_2(\frac{1}{\varepsilon})^{n/\alpha} \log\frac{1}{\varepsilon} \quad ;$$

c'est la seconde inégalité (2.2).

Preuve du théorème. Désignons par B' une boule de l'espace $\mathcal{C}_{n'}^{\alpha'}$, ayant les propriétés suivantes: 1. toute fonction de $\mathcal{C}_{n'}^{\alpha'}$ à support compact appartient à $\lambda B'$ pour λ assez grand 2. B' est contenu dans $L^\infty(\mathbb{R})$ et peut être recouvert par $N(\varepsilon)$ boules de $L^\infty(\mathbb{R})$ de rayon ε, avec $\log N(\varepsilon) = O((\frac{1}{\varepsilon})^{n'/\alpha'} \log\frac{1}{\varepsilon})$ 3. pour $\varphi \in B'$ on a

$$|\varphi(X_1,\ldots X_{n'}) - \varphi(Y_1,\ldots Y_{n'})| \leq \sup|X_j - Y_j|.$$

La condition 2. peut être satisfaite à cause de (2.2), et la condition 3. à cause de l'hypothèse $\alpha' \geq 1$. Observons qu'on peut se restreindre, dans les superpositions, à des fonctions de $C_{n'}^{\alpha'}$ à support compact, donc appartenant à la réunion des boules $\lambda B'$.

Désignons par $B'(s,\lambda)$ l'ensemble des fonctions représentables par superposition d'ordre s de fonctions appartenant à la boule $\lambda B'$, opérant sur $\cos t_1$, $\sin t_1, \ldots \cos t_n$, $\sin t_n$. Les éléments de $B'(s,\lambda)$ sont donc de la forme

$$\varphi(U_1,\ldots U_{n'}) \quad ((\varphi \in \lambda B', \quad U_j \in B'(s-1, \lambda)).$$

D'après la condition 2., toute $\varphi(U_1,\ldots U_{n'}) \in B'(s,\lambda)$ est approchable uniformément à moins de $\frac{\varepsilon}{2}$ près par des fonctions $\psi(U_1,\ldots U_{n'})$, où les ψ sont les centres de boules de rayon $\frac{\varepsilon}{2}$ recouvrant $\lambda B'$, le nombre des ψ étant $N(\frac{\varepsilon}{2\lambda})$. D'après la condition 3., les $\psi(U_1,\ldots U_{n'})$ sont approchables uniformement à $\frac{\varepsilon}{2}$ près par des fonctions $\psi(V_1,\ldots V_{n'})$, où les V_j sont des centres de boules de rayon $\frac{\varepsilon}{2\lambda}$ recouvrant $B'(s-1, \lambda)$. Le nombre minimum des fonctions $\psi(V_1,\ldots V_{n'})$ est

$$N(\frac{\varepsilon}{2\lambda})(N(\frac{\varepsilon}{2\lambda}, B'(s-1,\lambda))^{n'}.$$

Il résulte de la seconde inégalité (2.2), par récurrence sur s, que

$$H(\varepsilon, B'(s,\lambda)) \leq C(\frac{1}{\varepsilon})^{n'/\alpha'} \log \frac{1}{\varepsilon}$$

où $C = C(\alpha',n',\lambda,s)$. L'hypothèse $\frac{n'}{\alpha'} < \frac{n}{\alpha}$ et la première inégalité (2.2) entraînent

$$N(\varepsilon, B'(s,\lambda)) < N(\varepsilon,B^{\alpha})$$

pour ε petit, donc $B'(s,\lambda)$ n'est pas dense dans la boule B^{α}. Il en est de même pour toute boule dans $C^{\alpha}(T^n)$, donc la réunion des $\overline{B'(s,\lambda)}$ (s et λ entiers positifs) est un ensemble de première catégorie dans $C^{\alpha}(T^n)$. Cela achève la preuve du théorème.

III. THEOREME DE KOLMOGOROV (1957)

Sur la part prise par Arnold et Kolmogorov à la solution (négative) du 13ème problème de Hilbert nous renvoyons à [1]. Voici le théorème de Kolmogorov (1957).

Toute fonction continue réelle définie sur I^n ($I = [0,1]$, n _entier_ ≥ 2) _est représentable sous la forme_

$$f(x_1,\dots x_n) = \sum_{q=1}^{2n+1} g_q \left(\sum_{p=1}^{n} \varphi_{pq}(x_p) \right) \tag{3.1}$$

où les g_p _et les_ φ_{pq} _sont des fonctions continues d'une variable réelle, les_ φ_{pq} _sont strictement croissantes sur_ I, _et les_ φ_{pq} _ne dépendent pas de_ f.

Désignons par Γ_p la courbe d'equation

$$X_q = \varphi_{pq}(t) \quad (t \in I, \quad q = 1,2,\dots 2n+1).$$

Nous dirons que c'est un "arc croissant" dans $\mathbb{R}^{2n+1}$; cela veut dire que les coordonnées sont fonctions strictement croissantes du paramètre. Posons

$$E = \Gamma_1 + \Gamma_2 + \dots \Gamma_n \tag{3.2}$$

(somme algébrique). Le théorème de Kolmogorov dit que

1. E _n'a pas de point double_, c'est-à-dire que tout point de E

s'écrit d'une seule façon sous la forme

$$M = M_1 + M_2 + \ldots + M_n, \qquad M_p \varepsilon \Gamma_p.$$

2. E <u>est ensemble d'interpolation</u> au sens que toute fonction continue sur E s'écrit sous la forme

(3.3) $$g_1(X_1) + g_2(X_2) + \ldots + g_{2n+1}(X_{2n+1}).$$

Il existe plusieurs variantes du théorème de Kolmogorov, précisant des propriétés que l'on peut imposer aux fonctions φ_{pq} ou aux fonctions g_q. Voici l'une d'entre elles, due à Sprecher.

Soit $\lambda_1, \ldots \lambda_n$ des nombres strictement positifs tels que les 2^n sommes $\sum_1^n \varepsilon_p \lambda_p$ ($\varepsilon_p = 0$ ou 1) soient distinctes.* Alors on peut choisir $\varphi_{pq} = \lambda_p \varphi_q$. En d'autres termes, l'ensemble E est de la forme

(3.4) $$E = \lambda_1 \Gamma + \lambda_2 \Gamma + \ldots + \lambda_n \Gamma.$$

Dans (3.2), on peut se limiter aux arcs croissants Γ_p joignant les points $(0,0,\ldots 0)$ et $(\frac{1}{n}, \frac{1}{n}, \ldots \frac{1}{n})$, de façon que E soit contenu dans le cube I^n mais ne soit pas contenu dans un cube plus petit. De même, dans (3.4) on peut se limiter aux arcs croissants Γ joignant $(0,0,\ldots 0)$ et $(1, 1, \ldots 1)$. On peut alors, dans les formules (3.1) et (3.4), se restreindre à des fonctions g_q définies sur I. Lorentz a montré qu'on peut choisir $g_1 = g_2 = \ldots = g_{2n+1} (=g)$.

Les propriétés 1. et 2. de E sont des propriétés génériques. Si l'on munit l'ensemble des applications croissantes de I sur $\frac{1}{n} I$ de la topologie de la convergence uniforme, <u>quasi-sûrement (c'est-à-dire pour quasi tout choix des</u> φ_{pq}<u>) l'ensemble</u> E <u>donné par</u> (3.2)

* Dans [9] l'hypothèse indiquée sur λ_p est insuffisante; l'erreur est corrigée dans [1].

jouit des propriétés 1. *et* 2.: E *n'a pas de point double, et toute fonction continue sur* E *s'écrit*

(3.5) $g(X_1)+g(X_2)+\ldots+g(X_{2n+1}) \quad (g\in C(I))$.

Si l'on munit l'ensemble des applications croissantes de I sur I de la topologie naturelle, il en est de même pour l'ensemble E donné par (3.4) quasi-sûrement (c'est-à-dire pour quasi tout choix des φ_q). La démonstration de ce fait constitue la démonstration la plus facile du théorème de Kolmogorov (c'est la version, inspirée par les travaux de R. Kaufman et par le livre de G. Lorentz, que j'ai donnée dans mon cours à Orsay en 1967; voir aussi [3], [7], [8], [9], [1]).

Les propriétés 1. et 2. restent des propriétés génériques des ensembles E donnés par (3.2) et (3.4) si l'on considère les arcs croissants lipschitziens d'ordre α $(0<\alpha<1)$ au lieu d'arcs croissants arbitraires, ou même, si l'on impose aux φ_{pq} (ou aux φ_q) n'importe quelle majoration du module de continuité par une fonction $\omega(h)$ telle que $\lim_{h\to 0}\frac{\omega(h)}{h}=\infty$. Le fait que l'on puisse choisir les φ_{pq} lipschitziens (d'ordre 1) découle immédiatement de l'interprétation géométrique, parce que tout arc croissant admet une paramétrisation lipschitzienne; mais cela ne veut pas dire qu'il s'agisse d'une propriété générique.

On sait d'ailleurs, par le théorème de Vitushkin (1964) donné p.7 (partie I, dernière remarque), qu'on ne peut pas dans l'énoncé du thèorème de Kolmogorov choisir les φ_{pq} (indépendantes de f et) de classe C^1.

Nous allons maintenant nous préoccuper du choix des fonctions g.

IV. LES SOMMES $\Gamma_1 + \Gamma_2 + \ldots + \Gamma_n$ COMME ENSEMBLES D'INTERPOLATION

A quelles classes de fonctions g peut-on se restreindre dans (3.5) ?

La question présente de l'intérêt même dans le cas $n = 1$, et c'est dans ce cas qu'elle a d'abord été examinée [3]: on peut alors se restreindre à $g \in A(I)$, ensemble des restrictions à I des transformées de Fourier de fonctions sommables. Dans le cas $n > 1$, le résultat est inexact, mais nous allons établir l'énoncé de remplacement: <u>pour certains homéomorphismes ψ de I, que nous caractériserons, on peut prendre</u> $g \in (A \circ \psi)(I)$, <u>c'est-à-dire</u> $g = h \circ \psi$ <u>avec</u> $h \in A(I)$. Le premier énoncé dans ce sens était dû à R. Doss [10], qui choisissait $\psi(x) = e^x$.

Il peut être intéressant d'identifier I et $T = \mathbb{R}/\mathbb{Z}$, et de considérer $\dot{E}$ comme plongé dans T^{2n+1}. On peut alors choisir pour g une fonction valeurs-au-bord de fonction analytique dans le disque unité, comme l'a noté T. Hedberg [5]. Nous allons montrer, en application d'un théorème d'Alpar, qu'on peut choisir pour g une somme de série de Fourier-Taylor $\sum_0^\infty a_m e^{2\pi imt}$ uniformément convergente. On ne peut pas remplacer "uniformément" par "absolument".

Rappelons qu'on appelle ensemble de Helson dans un groupe abélien localement compact G un fermé E tel que l'algèbre $A(E)$ des restrictions à E des fonctions de $A(G)$ (transformées de Fourier des fonctions sommables sur le groupe dual) coincide avec $C(E)$, algèbre de toutes les fonctions continues sur E. Si on identifie I^m et T^m, on montre que les ensembles de Helson de $\mathbb{R}^m$ contenus dans I^m coincident avec les ensembles de Helson de T^m.

Une condition nécessaire pour qu'un compact $E \subset \mathbb{R}^m$ soit en ensemble de Helson est qu'il vérifie la "condition de maille": il contient au plus $K\,n\,s$ points dans toute maille de la forme $\sum_j^n \alpha_j M_j$, $\sum_1^n |\alpha_j| \leq 2^s$, où n et s sont arbitraires, $M_1, M_2, \ldots M_n$ sont des points arbitraires de $\mathbb{R}^m$ et K ne depend que de E([4], p. 34).

Il résulte immédiatement de la condition de maille que, si E contient la somme algébrique de deux ensembles infinis, E n'est pas un ensemble de Helson. Ainsi, pour $n > 1$, aucun ensemble $E = \Gamma_1 + \Gamma_2 + \dots \Gamma_n$ n'est un ensemble de Helson, lorsque les Γ_j sont des arcs croissants dans $\mathbb{R}^m$ (pour le moment, on ne suppose pas $m = 2n+1$).

Etant donné un homéomorphisme ψ de I (que nous supposerons croissant pour fixer les idées), désignons par $\psi(E)$ l'ensemble image de $E \subset \mathbb{R}^m$ par l'application

$$(4.1) \qquad \begin{array}{l} X_1 \to \psi(X_1) \\ X_2 \to \psi(X_2) \\ \vdots \\ X_m \to \psi(X_m). \end{array}$$

Supposons $E = \Gamma_1 + \dots + \Gamma$ et cherchons si, pour certains ψ, $\psi(E)$ est un ensemble de Helson.

Prenons d'abord pour ψ un polynôme de degré $\nu \leq n-1$. Considérons le polynôme à n variables $\psi(Y_1 + Y_2 + \dots + Y_n)$. Développé en somme de monômes, il s'écrit

$$(4.2) \qquad \psi(Y_1 + Y_2 + \dots Y_n) = \Sigma A_{j_1 \dots j_n} Y_1^{j_1} \dots Y_n^{j_n}$$

où $j_1 + j_2 + \dots + j_n \leq \nu$. Le nombre de termes dans la somme est inférieur à $(n+1)^\nu$. Choisissons sur chaque Γ_p N points $M_{p1}, M_{p2} \dots M_{pN}$, de telle façon que les N^n points

$$M_{1k_1} + \dots M_{nk_n} \qquad (k_p = 1, 2, \dots N)$$

soient tous distincts. Leurs images

$$M^{\psi}_{k_1..k_n} = \psi(M_{1k_1}+...+M_{nk_n})$$

par l'application (4.1) (aussi notée ψ) sont donc N^n points distincts. D'autre part chacune de leurs coordonées est de la forme (4.2). On peut donc écrire

(4.3) $$M^{\psi}_{k_1..k_n} = \Sigma\, M_{k_1..k_n j_1..j_n}$$

our par définition la q-ième coordonnées de $M_{k_1..k_n j_1..j_n}$ est $A_{j_1..j_n} Y_1^{j_1} ... Y_n^{j_n}$ lorsque la q-ième coordonée de M_{pk_p} est Y_p $(p = 1,2,.. n)$. Lorsque $j_1, j_2, ..., j_n$ et q sont fixés il y a moins de N^{ν} choix possibles pour $Y_1^{j_1} ... Y_n^{j_n}$. Il y a donc au total moins de $n^{\nu}N^{\nu}$ choix possibles pour $M_{k_1..k_n j_1..j_n}$, et la somme dans (4.3) contient moins de n^{ν} termes, tandis que (4.3) prend N^n valeurs, avec $n > \nu$. La condition de maille n'est donc pas satisfaite pour $\psi(E)$.

Il en est évidement de même si, au lieu de supposer que $\Gamma_1, ...\Gamma_n$ sont des arcs croissants, on suppose seulement que ce sont des ensembles infinis. Enonçons le résultat.

(4.4) <u>Si ψ est un polynôme de degré $\leq n-1$, l'image par (4.1) de la somme algébrique de n ensembles infinis n'est jamais un ensemble de Helson.</u>

Si E est une somme d'arcs croissants, $E = \Gamma_1 + \Gamma_2 + ... \Gamma_n$, contenue dans I^m et joignant les points $(0,0,... 0)$ et $(1, 1, ... 1)$, on peut se restreindre aux fonctions ψ définies sur I.

(4.5) <u>Si ψ coincide sur un sous-intervalle de I avec un polynôme de degré $\leq n-1$ $\psi(E)$ n'est pas un ensemble de Helson.</u>

Considérons maintenant l'ensemble E générique donné par (3.2), quand on suppose que les φ_{pq} sont des applications croissantes de I sur $\frac{1}{n}I$. Nous allons démontrer le résultat suivant.

(4.6) _Si ψ est un homéomorphisme de I qui ne coincide sur aucun sous-intervalle de I avec un polynôme de degré $\leq n-1$, quasi sûrement $\psi(E)$ est un ensemble de Helson. Plus précisément, toute fonction continue sur $\psi(E)$ s'écrit sous la forme_

(4.7) $$h(x_1)+h(x_2)+\ldots+h(x_{2n+1})$$

_où_ $h \in A(I)$.

La preuve repose sur la proposition suivante, très voisine de celle donnée en [9] p. 233 et qui se démontre de la même façon.

(4.8) _Soit $B(I)$ un espace de Banach contenu dans $C(I)$ et vérifiant l'hypothèse (H) : il existe $c>0$ tel que, pour tout $\delta>0$ il existe des ensembles finis $D_p \subset \frac{1}{n}I$ $(p=1,2,\ldots n)$ ayant au moins un point dans tout sous intervalle de $\frac{1}{n}I$ de longueur δ, tels que l'application $D_1\times D_2\ldots\times D_n \to D_1+D_2+\ldots D_n$ soit injective, et tels que pour toute fonction u de module 1 sur l'ensemble $D_1+D_2+\ldots D_n$ il existe $g\in B(I)$, $\|g\|_{B(I)}\leq c$ $\|g\|_{C(I)}=1$, $g=u$ sur $D_1+D_2+\ldots D_n$._

Alors on peut choisir dans (3.5) $g\in B(I)$.

Admettons cette proposition. On sait que si F est un ensemble fini rationnellement indépendant sur I, toute fonction de module 1 sur F est uniformément approchable par des exponentielles $e^{imt}|_F$ (théorème de Kronecker), et il en résulte que toute fonction de module 1 sur I s'écrit $\sum_1^\infty a_m e^{imt}$, $\sum_1^\infty |a_m| \leq 2$. En application de la proposition, on peut donc prendre $g \in (A\circ\psi)(I)$ dans (3.5) dès que l'homéomorphisme ψ vérifie la condition

(4.9) <u>pour tout</u> $\delta > 0$ <u>il existe des ensembles finis</u> $D_p \subset \frac{1}{n} I$ $(p = 1,2,\ldots n)$ <u>ayant au moins un point dans tout sous-intervalle de</u> $\frac{1}{n} I$ <u>de longeur</u> δ, <u>tels que l'application</u> $D_1 \times D_2 \ldots \times D_n \to D_1 + D_2 + \ldots D_n$ <u>soit injective, et que</u> $\psi(D_1 + D_2 + \ldots + D_n)$ <u>soit un ensemble rationnellement indépendant</u>.

Supposons que ψ ne vérifie pas (4.9). Pour tout entier $N > 0$ on a la situation suivante. Désignons par Ω l'ouvert de $\mathbb{R}^{nN}$ défini par les inégalités

$$\frac{k-1}{nN} < \xi_{pk} < \frac{k}{nN} \qquad (p = 1,2,\ldots n; \quad k = 1,2,\ldots N)$$

Quel que soit le point $(\xi_{pk}) \in \Omega$ il existe des entiers $m_{k_1 \ldots k_n}$ non tous nuls $(k_p = 1,2,\ldots N)$ tels que

$$(4.10) \qquad \Sigma\, m_{k_1 \ldots k_n} \psi(\xi_{1k_1} + \ldots + \xi_{nk_n}) = 0.$$

On utilise maintenant un argument classique[†]. Pour un choix donné des coefficients $m_{k_1 \ldots k_n}$ l'ensemble des $(\xi_{pk_p}) \in \Omega$ vérifiant (4.10) est un fermé. La réunion de ces fermés, pour tous les choix possibles des $m_{k_1 \ldots k_n}$ est Ω. Donc (Baire) l'un au moins de ces fermés a un intérieur non vide; choisissons les $m_{k_1 \ldots k_n}$ correspondants, et un pavé dans Ω où (4.4) ait lieu. Quitte à restreindre ce pavé, on peut remplacer dans (4.10) ψ par $\psi * \delta_\varepsilon$, où δ_ε est une fonction C^∞ de support $[-\varepsilon,\varepsilon]$ très petit. Choisissons un n-uple $k_1,\ldots k_n$ tel que $m_{k_1 \ldots k_n} \neq 0$, et dérivons (4.10) modifié, n fois successivement, par

---

[†] Introduit, á ma connaissance, par A. Beurling et H. Helson, et exposé par Y. Katznelson dans son livre "An Introduction to Harmonic Analysis", p. 217.

rapport à $\xi_{1k_1}, \xi_{2k_2}, \ldots \xi_{nk_n}$. On obtient

$$m_{k_1 \ldots k_n} (\psi * \delta_\varepsilon)^{(n)} (\xi_{1k_1} + \ldots + \xi_{nk_n}) = 0$$

c'est-à-dire $(\psi * \delta_\varepsilon)^{(n)} = 0$ sur l'image du pavé par l'application $(\xi_{pk}) \to \xi_{1k_1} + \ldots + \xi_{nk_n}$. Sur cet intervalle, $\psi * \delta_\varepsilon$ est un polynôme de degré $\leq n-1$. En faisant varier δ_ε on voit qu'il en est de même pour ψ.

En d'autre termes, si ψ n'est égal à un polynôme de degré $\leq n-1$ sur aucun sous-intervalle de I, ψ vérifie la condition (4.9). On peut donc choisir $g \in (A \circ \psi)(I)$ dans (3.5), et $h \in A(I)$ dans (4.7). Cela achève la démonstration de (4.6).

Les énoncés (4.5) et (4.6) répondent complètement au problème indiqué au début de cette partie: caractériser les homéomorphismes ψ de I tels qu'on puisse choisir dans (3.5) $g \in (A \circ \psi)(I)$.

Voici l'énoncé correspondant pour T au lieu de I. On identifie I et T, et on considère $E = \Gamma_1 + \Gamma_2 + \ldots \Gamma_n$ plongé dans T^{2n+1}. "Quasi sûrement" signifie "pour quasi tout choix des φ_{pq}". On désigne par $A^+(T)$ la classe des fonctions $h(t) = \sum_0^\infty \hat{h}_m e^{2\pi imt}$ avec $\sum_0^\infty |\hat{h}_m| < \infty$

(4.11) <u>Si ψ est un homéomorphisme de T qui ne coincide sur aucun intervalle avec un polynôme de degré $\leq n-1$, quasi-sûrement $\psi(E)$ est un ensemble d'interpolation dans T^{2n+1} au sens suivant: toute fonction continue sur $\psi(E)$ s'écrit</u>

$$h(X_1) + h(X_2) + \ldots + h(X_{2n+1})$$

<u>avec</u> $h \in A^+(T)$.

Si ψ viole l'hypothèse, $\psi(E)$ n'est même pas ensemble de Helson. Choisissons pour ψ une fonction de Möbius en $e^{2\pi it}$

$$\psi(t) = \frac{1}{2\pi i} \log \left(\frac{e^{2\pi it}-a}{1-a\, e^{2\pi it}} \right) \qquad 0 < a < 1.$$

Un théorème d'Alpár [11] dit que, si $h \in A^+(T)$

$$h \circ \psi(t) = \sum_0^\infty b_m e^{2\pi imt}$$

où la série du second membre est uniformément convergente, ce que nous notons $h \circ \psi \in U^+(T)$.

(4.12) <u>Quasi-sûrement,</u> E <u>est un ensemble d'interpolation</u> T^{2n+1} <u>au sens suivant: toute fonction continue sur</u> E <u>s'écrit</u>

$$g(X_1) + g(X_2) + \ldots + g(X_{2n+1})$$

<u>avec</u> $g \in U^+(T)$.

Si l'on considère T^{2n+1} comme la frontière distinguée du polydisque D^{2n+1}, cela entraîne que toute fonction continue sur E se prolonge à D^{2n+1} en une fonction holomorphe dont la série de Taylor converge uniformément dans D^{2n+1}.

V. UNE PROPRIETE TOPOLOGIQUE GENERIQUE DES SOMMES $\Gamma_1 + \Gamma_2 + \ldots \Gamma$

Soit φ_{pq} ($p = 1,2,\ldots n$; $q = 1,2,\ldots m$) des applications croissantes de I sur $\frac{1}{n} I$, Γ_p l'arc de $\mathbb{R}^m$ paramétré par les φ_{pq}, et $E = \Gamma_1 + \Gamma_2 + \ldots + \Gamma_n$.

(5.1) <u>Si</u> $m \geq 2n+1$ <u>quasi-sûrement l'application</u>

(5.2) $\Gamma_1 \times \Gamma_2 \ldots \times \Gamma_n \to \Gamma_1 + \Gamma_2 + \ldots \Gamma_n$

<u>est injective. Si</u> $m \leq 2n$, <u>quasi-sûrement elle n'est pas injective</u>.

La première partie a été énoncée dans la partie III sous la forme: E n'a pas de point double. Seul le cas $m = 2n+1$ était considéré, mais le cas $m \geq 2n+1$ en dérive aussitôt.

Démontrons la seconde partie. Posons $\varphi_p = (\varphi_{pq})$ $(q = 1,2,\ldots m)$: c'est la paramétrisation de Γ_p. Posons $\varphi = (\varphi_{pq})$ $(p = 1,2,\ldots n;$ $q = 1,2,\ldots m)$ et soit φ_o l'espace des φ, où la convergence est définie comme convergence uniforme pour chaque φ_{pq}. Si l'on convient que les φ_{pq} sont croissantes au sens large, Φ_o est un espace métrique complet. La partie de Φ_o correspondant à des φ_{pq} strictement croissantes est un G_δ dense dans Φ_o, soit Φ.

La première étape consiste à montrer que Φ contient un ensemble dense pour lequel (5.2) n'est pas injective; elle est valable pour tout m sans restriction. La seconde étape sera de montrer que Φ contient un ouvert dense pour lequel (5.2) n'est pas injective.

Soit Ω un ouvert dans Φ, et $a = (a_1,a_2,\ldots a_n)$ un point intérieur de I^n $(0 < a_p < 1;$ p $1,2,\ldots n)$. Il existe un $\delta > 0$ et des intervalles ouverts I_{pq} $(p = 1,2,\ldots n;$ $q = 1,2,\ldots m)$ tels que si, pour chaque couple (p,q), φ_{pq} est une application continue strictement croissante de l'intervalle $[a_p-\delta,\ a_p+\delta]$ dans I_{pq}, le système (φ_{pq}) se prolonge en une $\varphi \in \Omega$. Choisissons arbitrairement un point X_{pq} dans chaque I_{pq}, puis un point $Y_{pq} \in I_{pq}$ de façon que

$$(5.3) \qquad Y_{pq} = X_{pq} + \delta_p,\quad \delta_p \neq 0,\quad \sum_{p=1}^{n} \delta_p = 0.$$

Pour chaque p, choisissons $b_p \in [a_p-\delta,\ a_p+\delta]$ de façon que $\delta_p(b_p-a_p) > 0$. Soit A_p et B_p deux intervalles disjoints contenus dans $[a_p-\delta,\ a_p+\delta]$, respectivement centrés en a_p et b_p.

Pour chaque couple (p,q), soit φ_{pq} une application $A_p \cup B_p$ dans I_{pq} strictement croissante, affine sur A_p et affine sur B_p, telle que $\varphi_{pq}(a_p) = X_{pq}$ et $\varphi_{pq}(b_p) = Y_{pq}$. On vérifie que ces conditions sont compatibles. A cause de (5.3) on a

$$(5.4)\qquad \sum_{p=1}^{n} \varphi_{pq}(b_p) = \sum_{p=1}^{n} \varphi_{pq}(a_p) \qquad (q = 1,2,\ldots m).$$

On peut prolonger (φ_{pq}) en une $\varphi \varepsilon \Omega$. Pour ce φ l'application (5.2) n'est pas injective. Cela termine la première étape.

L'image de $A_1 \times A_2 \times \ldots \times A_n$ par l'application

$$(5.5)\qquad (x_1, x_2 \ldots x_n) \to \sum_{p=1}^{n} \varphi_p(x_p)$$

à valeurs dans R^m est un cube affine de dimension n, A. De même l'image de $B_1 \times B_2 \times \ldots B_n$ par (5.5) est un cube affine B, de dimension n, dans $\mathbb{R}^m$. Ces deux cubes ont un point commun d'après (5.4).

Nous n'avons pas encore utilisé l'hypothèse $m \leq 2n$. Supposons pour le moment $m = 2n$. Quitte à modifier les φ_{pq} sur les B_p, on peut supposer que A et B (de dimension n dans $\mathbb{R}^{2n}$) se coupent en un seul point.

Si φ varie continûment dans Ω, A et B se déforment continûment. La fin de la démonstration résulte donc du lemme suivant.

(5.6) <u>Si</u> A' <u>et</u> B' <u>sont deux cubes topologiques assez voisins de</u> A <u>et de</u> B <u>respectivement, on a</u> $A' \cap B' \neq \emptyset$.

Voici la preuve de (5.6), selon Albert Fathi et François Laudenbach (cas $m = 2n$). On peut définir le degré d'une application continue $f : S^k \to S^k$ (sphère de dimension k) comme le degré d'une application C^1 suffisamment proche de f.

Rappelons que le degré d'une application C^1 de S_k dans S_k se calcule en prenant la préimage d'une valeur régulière: c'est la différence entre le nombre de points "positifs" (où le déterminant de l'application dérivée est positif) et le nombre de points "negatifs". Le degré ne change pas par homotopie.

Soit D^n le disque unité de $\mathbb{R}^n$. Considérons $\mathbb{R}^{2n} = \mathbb{R}^n \times \mathbb{R}^n$, et j_1 (resp. j_2) l'injection de D^n dans $\mathbb{R}^{2n}$ comme disque unité du premier facteur (resp. du second facteur). Soit $h: D^n \times D^n \to \mathbb{R}^{2n}$ définie par

$$h(x,y) = j_1(x) - j_2(y).$$

Le fait que l'image de h contienne 0 traduit exactement le fait que les disques $j_1(D^n)$ et $j_2(D^n)$ ont un point commun.

Le bord $\partial(D^n \times D^n)$ est homéomorphe à S^{2n-1}, et $R^{2n}\setminus\{0\}$ a le type d'homotopie de S^{2n-1}. On peut donc parler du degré d'une application $\partial(D^n \times D^n) \to \mathbb{R}^{2n}\setminus\{0\}$. Par exemple, si les orientations sont bien choisies, on a

$$\deg(h \mid \partial(D^n \times D^n)) = +1.$$

(5.7) LEMME. <u>Soit</u> f_1 <u>et</u> f_2 <u>des approximations</u> C^0 <u>de</u> j_1 <u>et</u> j_2. <u>Alors</u> $f_1(D^n)$ <u>et</u> $f_2(D^n)$ <u>se rencontrent</u>.

Preuve. Soit $k(x,y) = f_1(x) - f_2(y)$. Cette application est une C^0 approximation de h. Donc $k(\partial(D^n \times D^n))$ est dans $\mathbb{R}^{2n}\setminus\{0\}$ et $\deg(k \mid \partial(D^n \times D^n)) = +1$. Si f_1 et f_2 n'ont pas de valeurs communes, k prend ses valeurs dans $\mathbb{R}^{2n}\setminus\{0\}$, donc $k|\partial(D^n \times D^n)$ est homotope à une application constante dans $\mathbb{R}^{2n}\setminus\{0\}$ (c'est k lui-même qui donne l'homotopie). Alors $\deg(k|\partial(D^n \times D^n) = 0$, ce qui est une contradiction.

Aux notations près, les énoncés (5.6) et (5.7) sont identiques.

Ainsi est établie la seconde partie de (5.1) quand $m = 2n$. Le cas $m \leq 2n$ s'en déduit immédiatement.

BIBLIOGRAPHIE

[1] VITUSHKIN, A.G. On representations of functions by means of superpositions and related topics. L'Enseignement Mathématique 23 (1977), 255-320.

[2] PROCEEDINGS of Symposia in Pure Mathematics, vol. 28. Mathematical developments arising from Hilbert problems. Voir en particulier: Hilbert, 13th problem, pp. 20-21 (part 1). G. Lorentz, the 13th problem of Hilbert, pp. 419-430 (part 2).

[3] KAHANE, J.-P. Sur les réarrangements de fonctions de la classe A. Studia Math. 31 (1968), pp. 287-293.

[4] KAHANE, J.-P. Séries de Fourier absolument convergentes. Ergebnisse der Mathematik, vol. 50. Springer-Verlag (1970).

[5] HEDBERG, T. Sur les réarrangements de fonctions de la classe A et les ensembles d'interpolation pour $A(D^2)$. C.R. Acad.Sc. Paris 270 A (1970), 1491-1494.

[6] HEDBERG, T. A result on interpolation sets, in Studies in Fourier Analysis. Inst. Mittag-Leffler, mai 1971, p. 8.

[7] HEDBERG, T. Continuous curves whose graphs are Helson sets. Chap. IV in Thin sets in harmonic analysis. Edited by L.A. Lindahl and F. Poulsen, Marcel Dekker 1971.

[8] HEDBERG, T. The Kolmogorof superposition theorem. Appendix II in Topics in Approximation Theory, by H.S. Shapiro. Lecture Notes 187, Springer 1971.

[9] KAHANE, J.-P. Sur le théorème de superposition de Kolmogorof. J. Approx. Theory 13 (1975), 229-234.

[10] DOSS, R. Representations of continuous functions of several variables. Amer.J.Math. 98 (1976), 375-383.

[11] ALPAR, L. Sur certaines transformées de séries de puissances absolument convergentes sur la frontière de leur cercle de convergence. Maghar Tud. Akad. Mat. Kutato Int. Közl. 6 (1961), 157-164.

[12] DOSS, R. On the representation of continuous functions of two variables by means of addition and continuous functions of one variable. Coll. Math. 10 (1963), 249-159.

Université de Paris-Sud
Equipe de Recherche Associée
au CNRS (296)
Mathématique (Bât. 425)
91405 Orsay Cedex

IVAŠEV MUSATOV IN MANY DIMENSIONS

by T.W. Körner, Trinity Hall, Cambridge

1. Introduction

The question we shall consider dates back over half a century to the discovery of Mensov that there exists a measure whose support has Lebesgue measure zero yet whose Fourier transform tends to zero at infinity. This fact in turn raises the question of how fast we can make the Fourier transform tend to zero while restricting the (closed) support to be of Lebesgue measure zero ?

One constraint is obvious. If μ is a measure on the circle $\underline{T} = \underline{R}/2\pi\underline{Z}$ and $\Sigma|\mu(n)|^2$ converges, then by the Riesz Fischer theorem μ is a (Lebesgue) L^2 function and so must either be zero or have for support a set of positive Lebesgue measure.

A succession of authors (Littlewood, Weiner and Winner, Shaeffer, Salem) found measures whose Fourier transforms dropped away almost as fast as the constraint above allowed. The search was essentially brought to an end by the following result of Ivasev Musatov [1].

<u>Theorem 1.1</u> Suppose that $\varphi(n)$ is a decreasing sequence such that

1) $\sum_{n=1}^{\infty} \varphi(n)^2$ diverges

2) $n\varphi(n)^2 \to 0$ as $n \to \infty$

3) $n^{1+\varepsilon}(\varphi(n))^2 \to \infty$ as $n \to \infty$ for all $\varepsilon > 0$

4) We can find an m such that $n^m\varphi(n)$ is an increasing sequence.

Then we can find a positive measure μ with support of Lebesgue measure 0 yet with

$$\hat{\mu}(n) = O(\varphi(|n|)) \text{ as } |n| \to \infty$$

In a series of papers [4], [5] and [6] I have succeeded in simplifying the conditions of Ivašev Musatov's theorem.

Theorem 1.2 Suppose that $\varphi(n)$ is a positive sequence such that

(A) $\sum_{n=1}^{\infty} \varphi(n)^2$ diverges

(B) There exists a $K < 1$ such that for all $n > 1$ we have
$K\varphi(n) \geq \varphi(r) \geq K^{-1}\varphi(n)$ whenever $2n \geq r \geq n$.

Then we can find a positive measure $\mu \neq 0$ with support of Lebesgue measure zero yet with

$$|\hat{\mu}(n)| = O(\varphi(|n|)) \text{ as } |n| \to \infty$$

The object of this talk was to give a description of the proof of Theorem 1.2 as given in [6]. It was, however, only a description, since I omitted some of the calculations and added an implicit extra condition

(C) $\varphi(n)$ is well behaved

at one or two points to simplify the discussion. If the reader comes across a gap in the reasoning, he should either use (C) or refer back to [6] where, I hope, the argument is fully laid out.

2. A Counter Example

The following example from [5] shows that condition (B) cannot be dispensed with entirely even when φ is convex.

Theorem 2.1 There exists a convcx positive sequence $\varphi(n)$ such that $\Sigma\varphi(n)^2$ diverges for all $\alpha > 0$ yet if μ is a non zero measure on $\underline{T}$ with $|\hat{\mu}(n)| = O(\varphi(|n|))$ as $n \to \infty$ it follows that supp $\mu = \underline{T}$.

The proof starts from the observation that if $\hat{\mu}(r) = 0$ for $|r| \geq N$ and $\sum_{r=-N}^{N} |\hat{\mu}(r) > 0$ then μ is a non zero trigonometric polynomial and so takes the value zero only finitcly often. There is a corresponding result if we only demand that $\hat{\mu}(r)$ be very small for $|r| \geq N$.

Lemma 2.2 Let $\eta > 0$, $L \geq 1$ and N be given. Then we can find an ε depending on η, L and N such that if $\mu \varepsilon M(\underline{T})$ and

$$(1) \quad L^{-1} \leq \sum_{r=-N}^{N} |\hat{\mu}(r)| \leq L$$

$$(2) \quad |\mu(r)| \leq \varepsilon \quad \text{for } |r| \geq N$$

then supp $\mu \cap I \neq \emptyset$ for each interval of length η.

Proof Suppose the result was false for some η, L and N. Then we could find a sequence μ_n of measures such that

$$(1) \quad L^{-1} \leq \sum_{r=-N}^{N} |\hat{\mu}_n(r)| \leq L$$

$$(2) \quad \sup_{|r| \geq N} |\hat{\mu}_n(r)| \to 0 \quad \text{as } n \to \infty$$

(3) supp $\mu_n \cap I_n = \emptyset$ for some interval I_n of length η. But it is clear that if we cover $\underline{T}$ by a finite collection of intervals J_p of length $\eta/4$, then at least one of the J_p lies in infinitely many of the I_n. Thus, by extracting a subsequence if necessary, we may replace (3) by

(3') supp $\mu_n \cap J = \emptyset$ for all n and some interval J.

Now conditions (1) and (2) show that there exist a subsequence $n(j) \to \infty$ and a distribution S such that $\mu_{n(j)} \to S$ in the distributional sense. We have automatically

$$(1)'' \quad L^{-1} \leq \sum_{r=-N}^{N} |\hat{S}(r)| \leq L$$

$$(2)'' \quad \hat{S}(r) = 0 \text{ for all } |r| \geq N$$

(3)" supp $S \cap J = \emptyset$.

Conditions (1)" and (2)" tell us that S is a non zero trigonometric polynomial and contradict condition (3)". The result follows.

We have thus shown that if a well behaved measure μ has

$|\hat{\mu}(r)| < \psi(|r|)$ and ψ suddenly decreases to a very small value from some point N onwards, then the support of μ must be well distributed round the circle (figure 1).

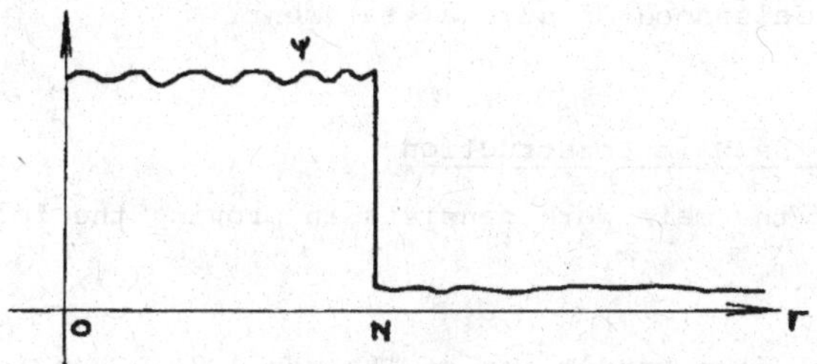

Figure 1

It seems fairly evident that by arranging for ψ to have an infinite series of steps with very sharp decreases (figure 2), we can force any

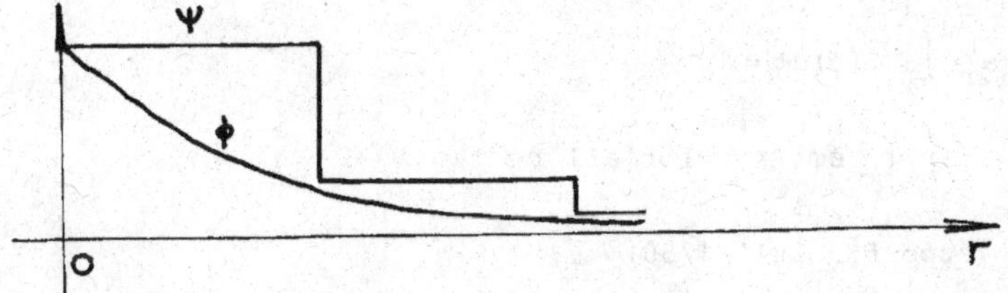

Figure 2

non zero measure μ with $|\hat{\mu}(r)| \leq \psi(r)$ to have support dense in $\underline{T}$ and so (since supp μ is closed) to have support the whole of $\underline{T}$. Admittedly ψ does not look very convex, but we can choose the length of each step to be as long as we want. In particular we can arrange to fit a φ "underneath ψ" such that

(a) φ is convex

(b) $(\Sigma\varphi(r)^n)$ diverges for each $n \geq 1$

(c) $\varphi(r)/\psi(r) \to 0$ as $r \to \infty$,

and this φ satisfies the conclusions of Theorem 2.1. The sceptical reader will find the details in §2 of [5].

At this point Varopoulos pointed out that the argument depends on supp μ being closed and so does not exclude the possibility of a result of the following form: – "If φ is convex and $\Sigma\varphi(r)^2$ diverges, then there exists a non zero measure μ and a set E of Lebesgue measure zero such that $|\mu|(E) = \|\mu\|$ yet $|\hat{\mu}(r)| \leq O(\varphi(|r|))$". I have no idea how to decide the truth or falsehood of such a statement.

3. Preliminaries to the Main Construction

It turns out that the main work consists in proving the following preliminary lemma.

Lemma 3.1 Let φ satisfy the conditions of Theorem 1.2. Then, given any $\varepsilon > 0$ we can find an infinitely differentiable function $f:\underline{T} \to \underline{R}$ such that

(i) $f(t) \geq 0$ for all $t \in \underline{T}$

(ii) $\frac{1}{2\pi} \int_{\underline{T}} f(t)\,dt = 1$

(iii) $|\hat{f}(r)| \leq \varepsilon\varphi(|r|)$ for all $r \neq 0$

(iv) $|\text{supp } f| \leq 2\pi(1-1/50)$

(iv)* supp f is "well distributed" round the circle.

(Here and for the rest of the talk $|E|$ means the Lebesgue measure of a set $E \subset \underline{T}$. The reader should ignore condition (iv)* for the time being; as the argument proceeds, it will become clear both what (iv) means and how easily it can be satisfied.)

On the surface Lemma 3.1 looks much easier to prove than Theorem 1.2 but if the reader reflects for a bit, he will see that this is not so. Before indicating how to prove Theorem 1.2 from Lemma 3.1, I should like to add some comments on this.

One way of trying to get a suitable function f for Lemma 3.1 is to start with a very smooth function as shown (figure 3).

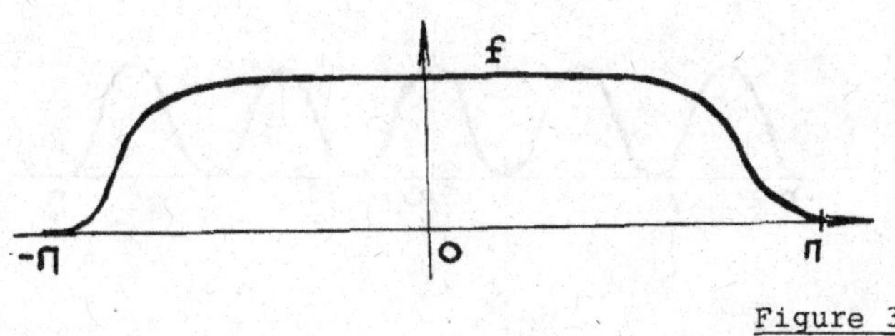

Figure 3

Since f is very smooth, we know that there is a function F with $F(r) \geq |\hat{f}(r)|$ which dies away very rapidly (figure 4). So far so good,

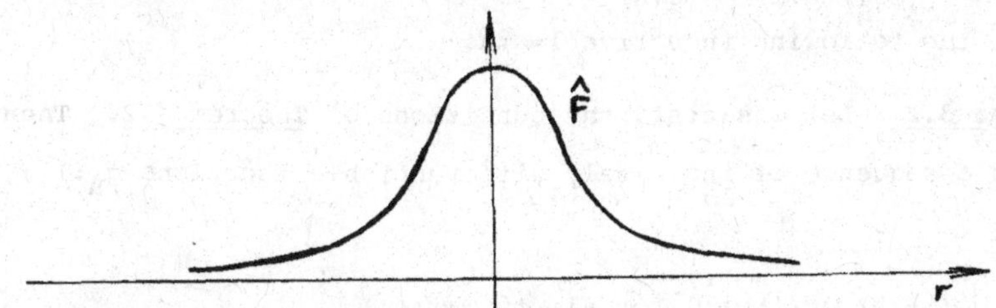

Figure 4

but we not only want $\hat{f}(r)$ to die away rapidly as $|r| \to \infty$ but to be small for all $r \neq 0$. In particular, if N in some "medium sized number", say 1000, then if f is a function of the type drawn in figure 3, at least one of the $\hat{f}(-N)$, $\hat{f}(1-N)$,..., $\hat{f}(-1)$, $\hat{f}(1)$, $\hat{f}(2)$,...,$\hat{f}(N)$ must be quite large. (This is pretty plausible, but may be proved by observing that, since f is smooth, elementary results on Cesaro summation show that

$\sup_{t\in T}|(N+1)^{-1}\sum_{n=-N}^{N}(N+1-|r|)\hat{f}(r)\exp irt - f(t)|$ is small and so looking at some t_o with $f(t_o) = 0$ one of the $\hat{f}(r)$ with $r \neq 0$ must be large.) Thus if ε is sufficiently small, the relation $|\hat{f}(r)| \leq \varepsilon\varphi(|r|)$ must fail for some r with $1 \leq |r| \leq N$.

Now suppose we concentrate on getting $\hat{f}(r)$ small for $|r| \leq N$. The obvious kind of function to use is something like that in figure 5, but the process of making f rough enough to make $|\hat{f}(r)|$ small for $|r| \leq N$ means that we can no longer rely on a smoothness argument to tell us

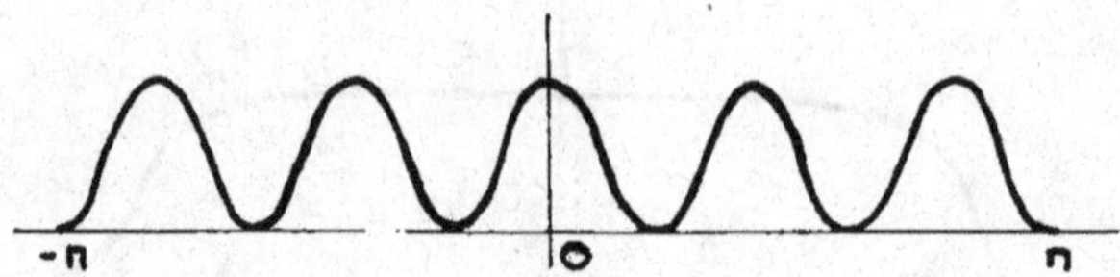

Figure 5.

that $\hat{f}(N+1)$, $\hat{f}(N+2)$,... are small.

How do we prove Theorem 1.2 from Lemma 3.1 ? Call the function of Lemma 3.1 f_ε. We claim that if $\varepsilon(j) \to 0$ fast enough, then the product $f_{\varepsilon(1)} f_{\varepsilon(2)} \dots f_{\varepsilon(n)}$ tends to a suitable μ. More formally we have the following inductive lemma.

Lemma 3.2 Let φ satisfy the conditions of Theorem 1.2. Then we can find a sequence of infinitely differentiable functions $g_n : \underline{T} \to R$ such that

$(i)_n$ $g_n(t) \geq 0$ for all $t \in \underline{T}$

$(ii)_n$ $2^{-1} + 2^{-n} \leq \frac{1}{2} \int_T g_n(t)\,dt \leq 2 - 2^{-n}$

$(iii)_n$ $|\hat{g}_n(r)| \leq (1-2^{-n})\varphi(|r|)$ for all $r \neq 0$

$(iv)_n$ $|\text{supp } g_n| \leq 2\pi(1-1/100)^n$

$(v)_n$ $\text{supp } g_n \subseteq \text{supp } g_{n-1}$

Proof. Suppose g_n has been constructed. We claim that setting $g_{n+1}(t) = g_n(t) f_\varepsilon(t)$ $[t \in \underline{T}]$ defines a suitable g_{n+1} provided only that ε is small enough. Most of the conditions on g_{n+1} are trivially verified. Since g_n and f_ε are infinitely differentiable positive functions so is g_{n+1}. Since $\text{supp } g_{n+1} = \text{supp } g_n \cap \text{supp } f_\varepsilon$ condition $(v)_{n+1}$ is also obviously satisfied. Moreover, since $\text{supp } f_\varepsilon$ is well distributed about the circle and covers less than $(1-1/50)^{\text{th}}$ of it, $\text{supp } f_\varepsilon$ can only cover less than about $(1-1/50)^{\text{th}}$ of $\text{supp } g_n$ and so certainly less than $(1-1/100)^{\text{th}}$ of it. Thus $|\text{supp } g_{n+1}| \leq (1-1/100)|\text{supp } g_n| \leq 2\pi(1-1/100)^{n+1}$.

The remaining two conditions $(ii)_{n+1}$ and $(iii)_{n+1}$ are thus the only non obvious ones. They will follow if we can show that

$$(ii)'_{n+1} \quad |\hat{g}_{n+1}(0) - \hat{g}_n(0)| \le 2^{-(n+1)}$$

$$(iii)'_{n+1} \quad |\hat{g}_{n+1}(r) - \hat{g}_n(r)| \le 2^{-n}\varphi(|r|) \quad [r \neq 0],$$

i.e. if we can show that $|\hat{g}_{n+1}(r) - \hat{g}_n(r)|$ is small for all r. But $\hat{g}_{n+1}(r) = (f_\varepsilon g_n)\hat{\ }(r) = \sum_{s=-\infty}^{\infty} \hat{f}_\varepsilon(r-s)\hat{g}_n(s) = \hat{f}_\varepsilon * \hat{g}_n(s)$. Now, as $\varepsilon \to 0$, $\hat{f}_\varepsilon(r) \to \hat{1}(r)$ (i.e. $\hat{f}_\varepsilon$ tends to the δ function on $\mathbb{Z}$), so since $\hat{g}_n(s)$ falls away rapidly, it is vaguely plausible that $\hat{g}_{n+1}(r)$ will be sufficiently close to $\hat{1} * \hat{g}_n(r) = \hat{g}_n(r)$ for all r.

To actually prove this, observe that conditions (B) of Theorem 1.2 implies $k^m\varphi(k) \to \infty$ as $k \to \infty$ for some $m \ge 1$, whilst the fact that g_n is infinitely differentiable implies $k^{m+1}\hat{g}_n(k) \to 0$ as $k \to \infty$. Thus for suitable constants $A_1, A_2, \ldots$ we have

$$\begin{aligned}
|\hat{g}_{n+1}(r) - \hat{g}_n(r)| &= |\sum_{s\neq r} \hat{f}_\varepsilon(r-s)\hat{g}_n(s) \\
&\le \sum_{s\neq r} \varphi(|r-s|)\ \hat{g}_n(s)| \\
&= \varepsilon \sum_{|r|/2>s} \varphi(|r-s|)\,|\hat{g}_n(s)| + \varepsilon \sum_{|s|\ge|r|/2,\, s\neq r} \varphi(|r-s|)\,|\hat{g}_n(s)| \\
&\le \varepsilon K \sum_{|r|/2>s} \varphi(|r|)\,|\hat{g}_n(s)| + g_n(s) + \varepsilon A_1 \sum_{|s|\ge|r|/2,\, s\neq 0} |\hat{g}_r(s)| \\
&\le \varepsilon K\varphi(|r|) \sum_{s=-\infty}^{\infty} |\hat{g}_n(s)| + \varepsilon A_2 \sum_{|s|\ge|r|/2,\, s\neq 0} s^{-(m+1)} \\
&\le \varepsilon K A_3 \varphi(|r|) + \varepsilon A_4\ |r|^{-m} \\
&\le \varepsilon A_5 \varphi(|r|),
\end{aligned}$$

which is what we want.

(The full argument is given in [6]. I found the idea of multiplying

f_ε and g_n in the original paper of Ivasev-Musatov [1]).

If we take $d\mu_n = g_n(t)dt$ then, since by (ii)$_n$ $||\mu_n|| \leq 4\pi$, we know that some subsequence of the μ_n converges weakly to a $\mu \varepsilon M(\underline{T})$. Such a μ will automatically inherit sufficient properties from the g_n to satisfy the conditions to Theorem 1.2.

4. Further Preliminaries

We have thus reduced the problem of proving Theorem 1.2 to that of proving Lemma 3.1. Our task is slightly simplified by observing that Lemma 3.1 is a consequence of an apparently less powerful result.

Lemma 4.1 There exists a constant C with the following property. Let φ satisfy the conditions of Theorem 1.2. Then we can find an infinitely differentiable function f such that

(i) $f(t) \geq 0$ for all $t \varepsilon \underline{T}$

(ii) $\frac{1}{2\pi}\int_T f(t)\,dt = 1$

(iii) $|\hat{f}(r)| \leq C\varphi(|r|)$ for all $r \neq 0$

(iv) $|\text{supp } f| \leq 2\pi(1-1/50)$

(iv)* supp f is "well distributed" round the circle.

Proof of Lemma 3.1 from Lemma 4.1 Observe that if φ satisfies the condition of Theorem 1.2 then so does $\psi = \varepsilon C^{-1}\varphi$. Applying Lemma 4.1 to ψ we obtain Lemma 3.1.

(At some point in the argument which follows the reader may ask whether condition (A) can be replaced by $\Sigma\varphi(n)^2 \geq 2\pi$. He should recall the proof just given and notice that whilst $\Sigma\varphi(n)^2$ divergent implies $\Sigma(\varepsilon C^{-1}\varphi(n))^2$ divergent, $\Sigma\varphi(n)^2 \geq 2\pi$ does not imply $\Sigma(\varepsilon C^{-1}\varphi(n))^2 \geq 2\pi$.)

We now come to the first novelty in the proof. This consists in replacing Lemma 4.1 by the following proposition.

Lemma 4.2 There exists a constant C with the following property. Let φ satisfy the conditions of Theorem 1.2. Then we can find an infinitely differentiable function g such that (i)' $1 \geq g(t) \geq -1$ for all $t \in \underline{T}$

(ii)' $\frac{1}{2\pi} \int_T g(t)\,dt = 0$

(iii)' $|\hat{g}(r)| \leq C\varphi(|r|)$ for all $r \neq 0$

(iv)' $|\{t: g(t) = -1\}| \geq 2\pi/50$

(iv)'* $|\{t: g(t) = -1\}|$ is well distributed round the circle.

Thus g looks something like this (figure 6):

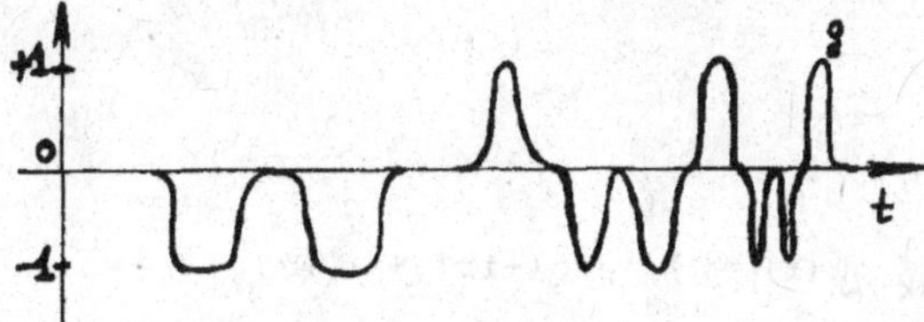

Figure 6

But if we set $f = 1 + g$ to give the function of figure 7, then the

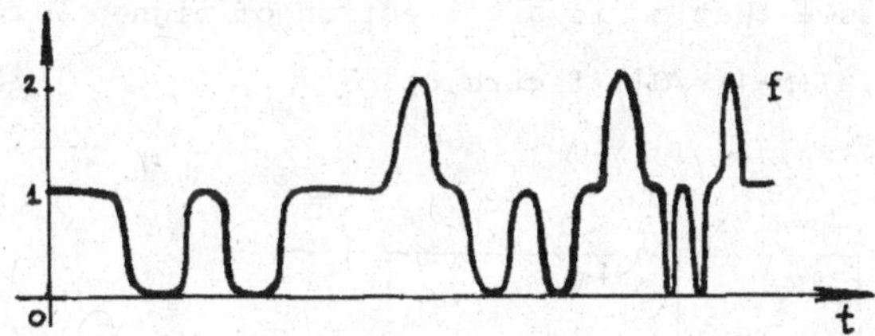

Figure 7

conditions on g given by Lemma 4.2 immediately imply the conditions on f given in Lemma 4.1.

5. The Basic Function

Although the preliminary manipulations just described take up a good half of the length of the proof, they contain no deep ideas. To make the proof work, we need something deeper. In Ivasev Musatov's proof (see [1] or my exposition [4]) the extra ingredient is the remarkable lemma of Van der Corput. In the proof given here, the extra ingredient is the remarkable theorem discovered by Shapiro and rediscovered by Rudin.

<u>Theorem 5.1 (Rudin Shapiro)</u> There exists a choice of + and - signs such that

$$\left|\sum_{k=0}^{N-1} \pm\exp(ikt)\right| \leq 10\sqrt{N} \quad \text{for all } t\in \underline{R}.$$

<u>Proof</u>. See <u>e.g</u>. [2] Chapter III, §6.

Now suppose $2\pi > \eta > 0$, $a \in \underline{T}$ and $N \geq 1$ are fixed for the time being. Taking + and - signs as in <u>Theorem 5.1</u>, we see that if

$$\sigma = \sum_{k=0}^{N-1} \pm\delta_{a+k\eta/N}$$

then
$$|\hat{\sigma}(r)| = \left|\sum_{k=0}^{N-1} \pm\hat{\delta}_{a+k\eta/N}(r)\right| = \left|\sum_{k=0}^{N-1} \pm\exp(-ir(a+k\eta/N))\right|$$

$$\left|\sum_{k=0}^{N-1} \exp(ik(-r\eta/N)\right| \leq 10\sqrt{N}.$$

Drawing diagrams, we see that σ is a collection of signed Dirac δ measures lying in $[a,a+(N-1)\eta/N]$ (figure 8).

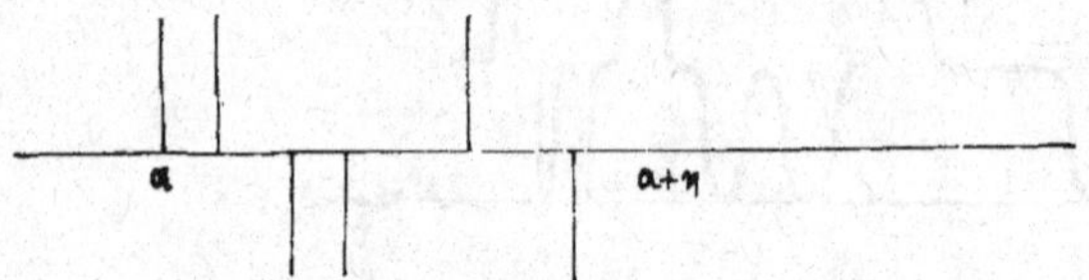

<u>Figure 8</u>

with $|\hat{\sigma}(r)| \leq F_1(|r|)$ where F_1 is the rather simple function drawn in figure 9.

<u>Figure 9</u>

We now proceed to modify this measure. Consider first the simple measures

$$\tau_1 = \delta_0 - \delta_{\eta/2N}$$
$$\tau_2 = \delta_0 - \delta_{\eta/4N}$$
$$\vdots$$
$$\tau \;= \delta_0 - \delta_{\eta/(2^m N)}.$$

We have $|\hat{\tau}_1(r)| = |1-\exp(-ir\eta/2N| = 2|\sin(r\eta/4N)|$ so $|\hat{\tau}_1(r)| \le G_1(|r|) = 2\max(1, |r|\eta/N)$. Notice that G drops away linearly as r goes to 0 (figure 10)

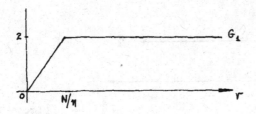

Figure 10

Similar considerations apply to $\tau_2, \tau_3, \ldots, \tau_m$ so that $|(\tau_1 * \tau_2 * \tau_3 * \ldots * \tau_m)^{\hat{}}(r)| = |\hat{\tau}_1(r)||\hat{\tau}_2(r)|\ldots|\hat{\tau}_m(r)| \le G(|r|) = 2^m(\max(1, (|r|\eta/N)^m)$. Now G drops away from its maximum value at N/η as fast as an mth power. If we now fix m large, G drops away very rapidly (as shown in figure 11) as

Figure 11

r goes from N/η to 0 and this is all we shall need.

If we consider $\tau = \sigma * \tau_1 * \ldots * \tau_m$ we see that τ is a collection of signed Dirac δ measures lying on all the points of the arithmetic progression $a + k\eta/2^m N$ where k runs from 0 to $2^m N-1$ (as shown in figure 12) with $|\tau(r)| \le F_2(|r|)$ where

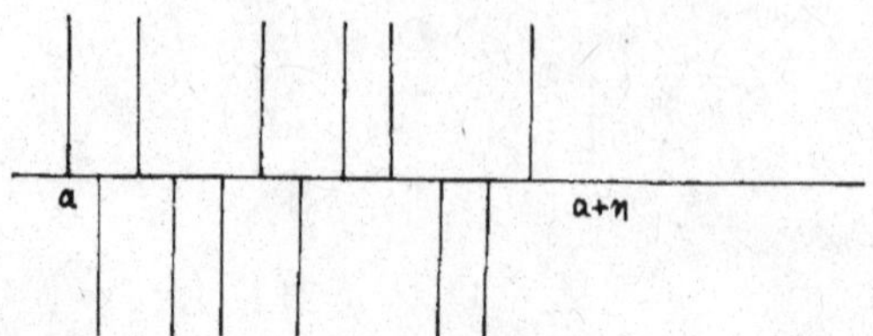

Figure 12

$F_2(|r|) = A_2\sqrt{N}(\max(1,(|r|\eta/N)^m)$. (In this discussion $A_2, A_3, \ldots$ will just be suitable constants.) We show F_2 in figure 13:

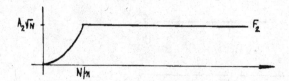

Figure 13

To make the second modification, we introduce the function h shown in figure 14. We demand that h be a well behaved infinitely

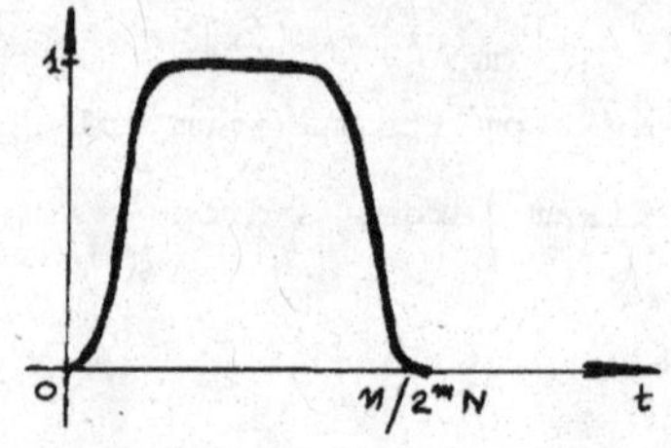

Figure 14

differentiable positive function such that

(a) $\operatorname{supp} h \subset |0, \eta/2^m N|$

(b) $0 \leq h(t) \leq 1$

(c) $|\{t: h(t) = 1\}| \geq (\eta/2^m N)/5$.

Since h is smooth we have (for each $M \geq 1$) $|\hat{h}(r)| \leq A_3 \int |h(t)| dt < \frac{|r|\eta}{2^m N} >^{-M}$ $[r \neq 0]$ where, provided we are sensible (and for example take h to be a scaled version of some fixed h_o, i.e. $h(t) = h_o(\lambda t)$ for $0 \leq t \leq \eta/2^m N$, $h(t) = 0$ with λ suitable) A_3 does not depend on η and N. Thus

$|\hat{h}(r)| \leq H(r)$ where $H(|r|) = A_4 \max(1, (|r|\eta/N)^{-M})/(N/\eta)$. If we now fix M large, H drops away very rapidly (see figure 15)

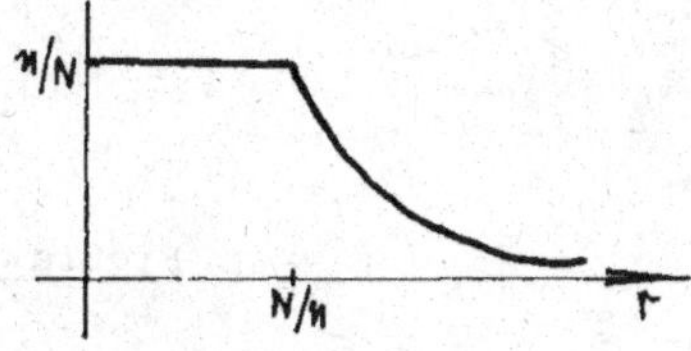

Figure 15

as r increases from N/η to 0.

Now set $v = \tau * h = \sigma * \tau_1 * \tau_2 * \ldots * \tau_m * h$. The resulting function (shown in figure 16) is automatically infinitely differentiable and has the

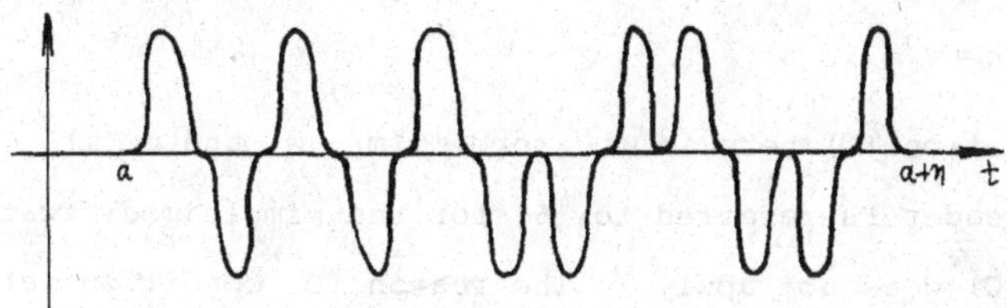

Figure 16

following properties:

(i)" $1 \geq v(t) \geq -1$ for all $t \epsilon \underline{T}$

(ii)" $\frac{1}{2\pi} \int_{\underline{T}} v(t)\,dt = 0$

(iii)" $|\hat{v}(r)| \leq V(|r|)$ for all $r \neq 0$

where $V(|r|) = H(|r|)F_2(|r|) = A_5 \eta N^{-1/2} \max\{|r|)\eta/N)^m, (|r| \eta/N)^{-M}\}$

(iv)" $|\{t : v(t) = -1\}| \geq \eta/10$

and (v)" $\operatorname{supp} v \subset [a, a+\eta]$.

We observe that V(r) attains its maximum value of $A_5 \eta N^{-1/2}$ at $r = N/\eta$ and falls away rapidly on either side of this value as shown in figure 17. The function v will be the basic component for our construction of g.

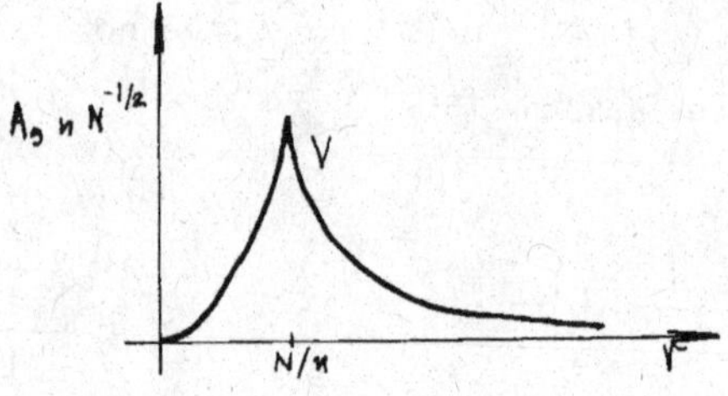

Figure 17

6. The Construction Completed

Let n be a reasonably large positive integer. Choose η and N so that

(a) $\eta = 2^n\varphi(2^n)^2$.

(b) $N/\eta = 2^n$

(We need condition (C) here in the form $m\varphi(m) \to \infty$ and $m\varphi(m)^2 \to 0$ as $m \to \infty$; the reader is referred to [6] for the simple modification required when (C) does not apply.) The reason for condition (a) will appear after a couple of paragraphs.

Now observe what happens to condition (iii)'. We now know that V(r) attains its maximum near 2^n with a maximum value of about $A_5\eta N^{-1/2} = A_5\eta^{1/2}(N/\eta)^{-1/2} = A_5(2^n\varphi(2^n)^2 \cdot 2^{-n})^{1/2} = A_5\varphi(2^n)$ and falls away rapidly on either side of this maximum as shown in figure 18.

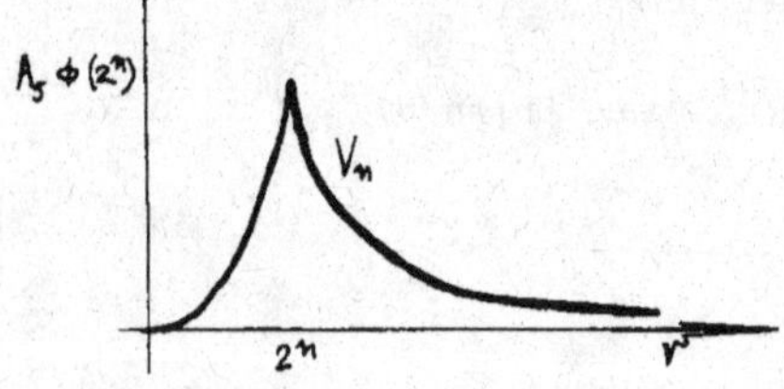

Figure 18

Introducing a suffix n we see that we have constructed an infinitely differentiable function V_n with

$(i)_n$ $1 \geq v_n(t) \geq -1$ for all t $\underline{T}$

$(ii)_n$ $\frac{1}{2\pi}\int_{\underline{T}} V_n(t)\,dt = 0$

$(iii)_n \quad |\hat{v}_n(r)| \leq V_n(|r|)$ where v_n has a maximum of $A_6\varphi(2^n)$ at 2^n and dies away rapidly as r goes to 0 or to ∞.

$(iv)_n \quad |\{t: v_n(t) = -1\}| \geq \eta_n/10$

and $(v)_n \quad \text{supp } v_n \subset |a_n, a_n + \eta_n|$.

Now, by condition A, $\eta_n = 2^n\varphi(2^n) \geq K^{-1} \sum_{s=2^n}^{2^{n+1}-1} \varphi(s)$ so $\sum_{n=1}^{\infty} \eta_n$ diverges.

Making use of condition (C) to suppose that $\eta_n \to 0$, this tells us that there exist N(1) and N(2) such that $\eta_n \leq 1/10$ for all $n \geq N(1)$ and $4\pi/5 \geq \sum_{n=N(1)}^{N(2)} \eta_n \geq 2\pi/5$. (This is the reason for condition (a) above.) We can thus make the intervals $[a_{n(1)}, a_{N(1)}+\eta_{N(1)}]$, $[a_{N(1)+1}, a_{N(1)+1}+\eta_{N(1)+1}]$, ... $[a_{N(2)}, a_{N(2)}+\eta_{N(2)}]$ disjoint and "well distributed" round the circle. Setting $g = \sum_{n=N(1)}^{N(2)} v_n$ we see that, since the supports of the v_n are disjoint,

(i)' $\quad 1 \geq g(t) \geq -1$ for all $t \in \underline{T}$

(ii)' $\quad \frac{1}{2\pi}\int g(t)\,dt = 0$

(iv)'* $\quad |\{t: g(t) = -1\}$ is well distributed round the circle.

Finally we estimate $\hat{g}(r)$. We have

$$|\hat{g}(r)| \leq \sum_{n=N(1)}^{N(2)} |\hat{v}_n(r)| \leq \sum_{n=N(1)}^{N(2)} V_n(|r|).$$

But we know that the $V_n(r)$ are sharply peaked as shown in figure 19.

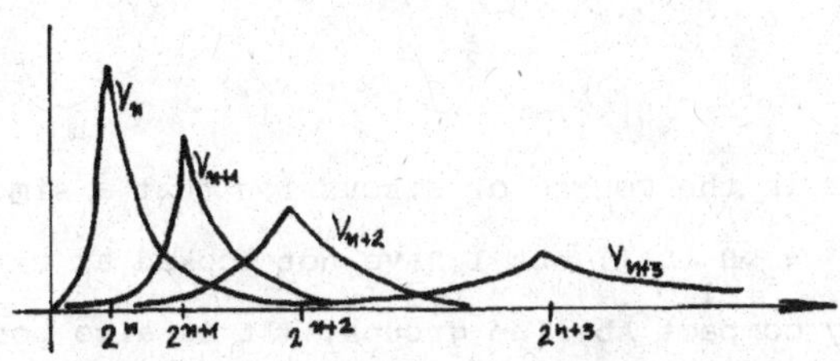

Figure 19

Thus if $2^m \leq |r| \leq 2^{m+1}$ the main contribution to the sum $\sum_{n=N(1)}^{N(2)} V_n(|r|)$ comes from the terms $V_m(|r|)$ and $V_{m+1}(|r|)$.

Thus $|\hat{g}(r)| \leq A_7(\varphi(2^m)+\varphi(2^{m+1}))$. But by condition (A) $\varphi(2^m), \varphi(2^{m+1}) \leq K\varphi(|r|)$, so we have

(iii)' $|\hat{g}(r)| \leq C\varphi(|r|)$.

(A more detailed estimate is given in [6]).

We have now proved Lemma 4.2 and so Theorem 1.2.

7. Comments

It is worth repeating that Ivasev Musatov's original result covers all "naturally occurring" φ such as $\varphi(n) = n^{-1/2}$, $\varphi(n) = (n \log n)^{-1/2}$, $\varphi(n) = (n \log n \log \log n)^{-1/2}$ and so on. However, I think that the proof presented here is simpler in some ways than that of Ivasev Musatov or my exposition of his method in [4]. (It is certainly much simpler than my attempt in [5] to push the method still further.) It also has the advantage over the previous method that it extends to several dimensions.

Theorem 1.2' Suppose $\varphi : \underset{\sim}{Z}^m \to \underset{\sim}{R}^+$ is such that

(A) $\Sigma\varphi(\underset{\sim}{n})^2$ diverges

(B) There exists a $K \geq 1$ such that for all $\underset{\sim}{n} \neq 0$ we have

$K\varphi(\underset{\sim}{n}) \geq \varphi(\underset{\sim}{n}) \geq K^{-1}\varphi(\underset{\sim}{n})$ whenever $2||\underset{\sim}{n}|| \geq ||r|| \geq ||\underset{\sim}{n}||$. Then we can find a positive measure $\mu \neq 0$ with support of Lebesgue measure zero yet with

$$|\mu(\underset{\sim}{n}) = O(\varphi(||\underset{\sim}{n}||)).$$

(Here $||\underset{\sim}{n}|| = \sqrt{(\sum_{i=1}^{m} n_i^2)}$.)

Gundy pointed out in the course of discussion that a similar result will hold for $D_\infty = \prod_{r=1}^{\infty} \{0,1\}$ but I have not looked at extensions to more general locally compact Abelian groups. It is also possible to modify condition (B) (which, as it stands, essentially imposes spherical symmetry on φ) but I cannot think of any useful modification.

There is one last mathematical remark that I want to make. If the reader does not-find it useful (or even comprehensible), he should not worry. I see the function f of Theorem 1.2 as a kind of comb whose teeth get fine and finer (see figure 20). If we want to estimate $\hat{f}(r)$

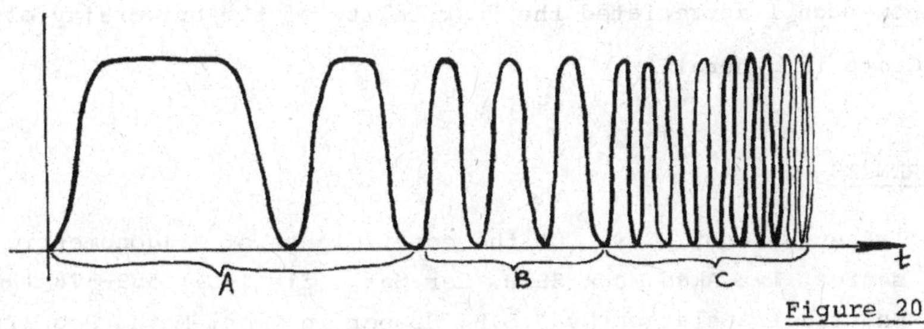

Figure 20

we have

$$\hat{f}(r) = \int_A f(t)\exp(-irt)\,dt + \int_B f(t)\exp(-irt)\,dt + \int_C f(t)\exp(-irt)\,dt$$

where f(t) varies very slowly compared with exp(-irt) on A, and f(t) varies very fast compared with exp(-irt) on A, and f(t) varies very fast compared with exp(-irt) on C. It "follows" that the contribution to $\hat{f}(r)$ from the A bit and the C bit are negligible. Thus $\hat{f}(r) = \int_B f(t)\exp(-irt)\,dt$. We now obtain an estimate of $\int_B f(t)\exp(-irt)\,dt$ of the form $\int f(t)\,dt \times$ (a cunning estimate based on the form of f).

Since we want $\frac{1}{2\pi}\int_{\underline{T}} f(t)\,dt = 1$ and since taking $\int f$ small over one interval would have to be compensated by taking it large over another, we might as well take $\int_B f(t)\,dt = \frac{B}{2\pi}$. Thus $|\hat{f}(r)| \leq \frac{B}{2\pi} \times$ (a cunning estimate based on the form of f). The length of |B| is the length of the interval over which the variation of f is comparable with exp(-irt), i.e. f has a "wavelength" between r/2 and 2r (or, if for convenience we study r of the form 2^n, between 2^{n-1} and 2^{n+1}). This, together with the demand that f be defined over the whole of $\underline{T}$ so that the "B_n" cover the whole of the circle, leads to conditions like (a) and (b) of Section 6.

The ideas just discussed can be extended to give results on Hausdorff measures. This is done in [6]. An interesting but complicated extension which shows that the μ in Theorem 1.2 may be chosen to be a translational measure (mésure de translation [3]) provided $n\varphi(n)$ is an increasing sequence tending to ∞ is given in [7].

I should like to end by saying how much I enjoyed this Conference and how much I appreciated the hospitality of the University of Crete and Crete in general.

REFERENCES

[1] Ivasev-Musatov, O.S. On the coefficients of trigonometric nul series, Ivz.Akad.Nauk SSSR, Ser.Mat., 21 (1957) 559-578 (Russian). English translation by J.L.B. Cooper in Amer. Math. Soc. Translations, Series 2, 14 (1960) 289-310.

[2] Kahane, J.-P. Séries de Fourier absolument convergentes. Springer Verlag, Berlin 1970.

[3] Kahane, J.-P. and Salem, R. Ensembles parfaits et séries trigonométriques. Hermann, Paris 1963.

[4] Körner, T.W. On the theorem of Ivašev-Musatov I. Ann. Inst. Fourier (Grenoble) 27 (1977) fasc. 3, 97-115.

[5] Körner, T.W. On the theorem of Ivašev-Musatov II. To appear in Ann. Inst. Fourier (Grenoble).

[6] Körner, T.W. On the theorem of Ivašev-Musatov III. Submitted for publication.

[7] Körner, T.W. On the theorem of Ivašev-Musatov IV. Submitted for publication.

BEMERKUNGEN ÜBER LINKSIDEALE IN GRUPPENALGEBREN

von Horst Leptin in Bielefeld

Bemerkungen über Linksideale in Gruppenalgebren

In verschiedenen Arbeiten ist in den letzten Jahren die Untersuchung der Struktur der Faltungsalgebra $L^1(G)$ für nicht kommutative und nicht kompakte, lokal kompakte Gruppen G aufgenommen. Dabei spielte das Studium der idealtheoretischen Struktur dieser Algebren eine besondere Rolle. Eines der Leitmotive, formuliert etwa in meiner Arbeit [3] ist die Frage nach dem Gültigkeitsbereich von Aussagen, als deren Modell man den klassischen Satz von Wiener-Godement-Segal über maximale Ideale in den Algebren lokal kompakter abelscher Gruppen betrachtet. Es liegt auf der Hand, daß für die Übertragung dieses Satzes auf Gruppen, die nicht mehr kommutativ oder kompakt sind, eine Vielzahl wesentlich verschiedener sinnvoller Möglichkeiten sich anbietet, weil nämlich - zum Beispiel - die Rolle der maximalen abgeschlossenen Ideale bei kommutativen $L^1(G)$ im allgemeinen Fall durch völlig verschiedene Objekte übernommen werden kann. So hat man zunächst die Wahl zwischen zweiseitigen oder einseitigen Idealen. Bei den zweiseitigen kann man dann primitive Ideale, "unitäre" primitive Ideale, maximale, maximale modulare Ideale usw. als elementare Objekte betrachten, vgl. etwa [5].

In [7] habe ich systematisch Fragen der harmonischen Analyse auf der Basis einseitiger Ideale untersucht. Als <u>einseitige Wienersche Eigenschaft</u> der Algebra $L^1(G)$ sei die Tatsache bezeichnet, daß jedes echte abgeschlossene Linksideal in $L^1(G)$ von einem stetigen nicht trivialen positiven Funktional annuliert wird. Hierzu ist offensichtlich äquivalent, daß jeder w*-abgeschlossene Links-G-Untermodul von $L^\infty(G)$ positive definit von Null verschiedene Funktionen enthält.

In [7], Theorem 10, habe ich gezeigt, daß eine zusammenhängende Gruppe G diese Eigenschaft genau dann besitzt, wenn sie direktes Produkt einer kompakten und einer abelschen Gruppe ist. Der nicht triviale Teil des Beweises besteht natürlich im Nachweis der Existenz von Linksidealen in $L^1(G)$, die von keinen positiven Funktionalen annuliert werden, sofern G nicht direkt in einen kompakten und einen abelschen Teil zerfällt. In diesem Fall läßt sich das Problem auf einen Typ von Algebren reduzieren, die wir intensiv in mehreren Arbeiten studiert haben [3], [6], [8], nämlich die "verschränkten Produkte" $\mathcal{L} = L^1(G,A)$ mit einer Banachschen G-Unteralgebra A aus $C_\infty(G)$, den stetigen komplexwertigen, im Unendlichen verschwindenden Funktionen auf G. Für diese läßt sich der Verband der abgeschlossenen Linksideale beschreiben durch die abgeschlossenen Unterräume einer gewissen Segal-Algebra A_1 in A. Die maximalen Linksideale entsprechen dabei den abgeschlossenen Unterräumen der Codimension 1 in A_1, d.h. im wesentlichen den stetigen nicht trivialen linearen Funktionalen auf A_1. Da A_1 stetig in jeden Raum $L^p(G)$, $1 \leq p \leq \infty$, eingebettet ist, definieren insbesondere die Funktionen $\omega \in (L^p)' = L^1(G)$, $1 \leq q \leq \infty$, stetige Funktionale auf A_1 und somit maximale abgeschlossene Linksideale $\mathcal{F}_\omega \subset \mathcal{L}$. Ein Linksideal aus wird nun genau dann von einem positiven Funktional annulliert, wenn es in einem F_ω enthalten ist für ein $\omega \in L^2(G)$. Es folgt hieraus insbesondere, daß in jeder Algebra $L^1(G)$, G zusammenhängend und nicht direktes Produkt abelscher und kompakter Gruppen, sogar maximale abgeschlossene Linksideale existieren, die nicht von positiv definiten Funktionalen annulliert werden.

Für eine große Klasse auflösbarer Liescher Gruppen läßt sich leicht zeigen, daß von positiven Funktionalen annullierte Linksideale in maximalen abgeschlossenen Linksidealen enthalten sind. Das führt auf die Frage, ob es möglicherweise abgeschlossene Linksideale in $L^1(G)$ gibt, die keine maximalen Oberideale besitzen. Für die Bewegungsgruppe M_2 der Ebene hat nun in der Tat Y. Weit zeigen können, daß solche Ideale existieren [9]. Wir werden hier zeigen, daß mit den in [7] entwickelten Methoden auch dieses Problem behandelt werden kann.

Ein anderes Problem tritt in der Theorie der Faltungshalbgruppen auf [2]: Die Untersuchung hypoelliptischer invarianter

Differentialoperatoren führt auf Faltungshalbgruppen $\{p_t\}_{t>0}$ in $L^1(G)$, z.B. für die Heisenberggruppen H_k. Die Frage nach Sätzen vom Wiener-Tauberschen Typ stellt einen dann vor das Problem, ob ein p_t, $t>0$, ein echtes Linksideal erzeugen kann. Rechnet man modulo primitiven Idealen, d.h. im wesentlichen in den Algebren $L^1(G,Q)$, so erkennt man, daß dieses Problem mit einer eigentümlichen Eigenfunktionen-Entwicklung hermite'-scher Funktionen aus $L^1(G,Q)$ zusammenhängt.

Wir werden in diesem Aufsatz keine abschließenden neuen Sätze beweisen, wir wollen lediglich zeigen, wie die u.a. in den Arbeiten [6], [7], [8] dargestellten Methoden und Begriffe auch auf die oben beschriebenen Probleme mit Nutzen angewandt werden können.

I.

Die $(2k+1)$-dimensionale Heisenberggruppe H_k läßt sich auffassen als halbdirektes Produkt

$$H_k = \mathbb{R}^k \propto (\mathbb{R}^k \times \mathbb{R})$$

des k-dimensionalen Raumes $V=\mathbb{R}^k$ mit dem k+1-dimensionalen Raum $N=V\times\mathbb{R}$ als normaler Untergruppe. Bezeichnet xy für $x,y\in V$ das übliche innere Produkt auf V, so läßt sich die Wirkung von V auf N durch $(x,t)\mapsto(x,t)^g=(x,-gx+t)$ beschreiben, $x,g\in V$, $t\in\mathbb{R}$. Für $L^1(H_k)$ erhalten wir

$$L^1(H_k) \cong L^1(V,L^1(V\times\mathbb{R})) \cong L^1(V,A(V\times\mathbb{R}))$$

mit der Fourier-Algebra $A=A(V\times\mathbb{R})$ von $N=V\times\mathbb{R}$. Die Wirkung von $g\in V$ auf Funktionen $a\in A$ ist dabei durch die Formel

$$a^g(x,t) = a(x+tg,\ t)$$

bestimmt.

Die Fourier-Algebra $A=A(V\times\mathbb{R})$ ist also eine V-Algebra, ebenso wie die Fourier-Stieltjes Algebra $B=B(V\times\mathbb{R})$. Die Algebra B läßt sich mit der adjungierten Algebra A^b von A identifizieren [4] und außerdem als Teil der adjungierten Algebra L^b von $L = L^1(V,A)$ auffassen. Die Algebra B^V der V-Invarianten aus B liegt dabei im Zentrum von L^b. Es ist klar, daß $u \in B$ in B^V liegt, wenn $u = u(x,t)$ von x nicht abhängt, man kann infolgedessen B^V mit $B(\mathbb{R})$ identifizieren, insbesondere liegt $A(\mathbb{R})$ in B^V.

Das Ideal $A_o = \{a \in A;\ a(x,o) = o,\ x \in V\}$ ist V-invariant und $L_o = L^1(V,A_o)$ ist ein abgeschlossenes zweiseitiges Ideal in L mit $L/L \cong L^1(V\times V) \cong L^1(\mathbb{R}^{2k})$. Es ist ersichtlich, daß die "nicht kommutativen" Probleme der Idealtheorie von L sich im Ideal L_o wiederspiegeln. Hier ist eine weitere Reduktion möglich: Sei z.B. Λ ein abgeschlossenes Linksideal in L_o. Zunächst ist $L_o = (L_o^+ \oplus L_o^-)^-$ die Hülle der idealtheoretischen direkten Summe der Ideale

$$L_o^{\pm} = L^1(V,A_{\pm}) \quad \text{mit} = A_{\pm} = \{a \in A;\ a(x,t) = o \ \text{ für } \pm t \le o\}.$$

Es folgt leicht, daß $\Lambda = (\Lambda^+ \oplus \Lambda^-)^-$ mit $\Lambda^{\pm} = (L_o^{\pm}\Lambda)^-$. Ist also $\Lambda \neq L_o$, so folgt etwa $\Lambda^+ \neq L_o^+$ und die Frage z.B. nach maximalen Oberidealen von Λ ist auf die entsprechende Frage in L_o^+ reduziert. Hier können wir das Problem wie folgt noch weiter reduzieren:

Ist C das Ideal aller $u \in A(\mathbb{R})$ mit kompakten Trägern in $\mathbb{R}^+ = \{t > o\}$, so ist $C \subset B^V$ und $\overline{AC} = A_+$, also $(L_o C^{\#})^- = (L_o^+ C^{\#})^- = L_o^+$ [4], folglich existiert ein $b \in C$, etwa mit Träger supp $b \subset [1,2]$, mit $L_o^+ b^{\#} \not\subset \Lambda$. Ist dann

$$I = \{a \in A_+;\ a(x,t) = o,\ 1 \le t \le 2\},$$

so folgt $(\Lambda + L^1(V,I))^- \neq L_o^+$, da $L^1(V,I)b^{\#} = 0$. Somit ist der Abschluß des Bildes von Λ in der Faktoralgebra

$$L_o^+/L^1(V,I) \cong L^1(V,A_+/I)$$

ein eigentliches Linksideal.

Sei $Z = [1,2]$ und $X = V\times Z$. Wir betrachten X als lokal kompakten V-Raum mit der Wirkung

$g(y,z) = (g+y,z)$

für $g,y \in V$, $z \in Z$. Die Bahnen sind dann die Mengen $X_z = V \times \{z\}$ und der Bahnenraum ist gleich Z. Für $f \in A_+$ sei

(1) $\tilde{f} : (y,z) \mapsto \tilde{f}(y,z) = f(zy,z)$.

Dann ist $\tilde{f} \in C_\infty(X)$ und $\tilde{f^v} = \tilde{f}^v$ mit $h^v(x) = h(vx)$ für Funktionen h auf X. Somit definiert $f \mapsto \tilde{f}$ einen V-Isomorphismus der V-Algebra A_+/I auf eine invariante Unteralgebra $\tilde{A}$ aus $C_\infty(X)$.

Als nächstes betrachten wir die Bewegungsgruppe M_n des $\mathbb{R}^n$. Ist $G = SO(n,\mathbb{R})$ die spezielle orthogonale Gruppe des $\mathbb{R}^n$, so ist $M_n = G \propto \mathbb{R}^n$ das halbdirekte Produkt von G und $\mathbb{R}^n$, mit der kanonischen Wirkung von G auf $\mathbb{R}^n$. Mit der Fourier Algebra $A = A(\mathbb{R}^n)$ erhalten wir dann

$$L^1(M_n) \cong L^1(G, L^1(\mathbb{R}^n)) \cong L^1(G,A)$$

und auch hier lassen sich idealtheoretische Fragen über $L^1(M_n)$ in der Regel wieder auf das Ideal

$$\mathcal{L}_o = L^1(G, A_o(\mathbb{R}^n))$$

mit $A_o(\mathbb{R}^n) = \{a \in A;\ a(o) = o\}$ reduzieren. Die Algebra A_o^G der G-invarianten Funktionen aus A_o besteht hier aus den radialen Funktionen $f \in A_o$, für die also $f(x) = f(r)$ mit $r = |x|$ für alle $x \in \mathbb{R}^n$ gilt. Diese Algebra A_o^G läßt sich als Unteralgebra von $C_\infty(\mathbb{R}^+)$, den stetigen, in 0 und ∞ verschwindenden komplexwertigen Funktionen betrachten und ebenso wieder als Teil des Zentrums der adjungierten Algebra $\mathcal{L}_o^b$ von $\mathcal{L}_o$. Da auch hier $(A_o A_o^G)^- = A_o$ gilt und die Funktionen aus A_o^G mit kompakten Trägern in $\mathbb{R}^+$ in A_o^G ein dichtes Ideal bilden, erkennt man genau wie im Fall der Heisenberggruppen, daß die kritischen Fragen sich auf die Untersuchung der Algebra $L^1(G, A_o/K)$ reduzieren lassen, in der

$$K = \{a \in A_o;\ a(x) = o \text{ für } 1 \leq |x| \leq 2\}$$

ist.

Wir bezeichnen mit e den Einheitsvektor $\{1,o,\ldots,o\}$ aus $\mathbb{R}^n$ und mit F die Fixgruppe von e in G. Es ist also $F \cong SO(n-1,\mathbb{R})$ und $gF \mapsto ge$ ist ein Diffeomorphismus des homogenen Raumes G/F auf die $(n-1)$-Sphäre S^{n-1}. Setzen wir $Y = G/F$, $Z = \{1,2\}$ und $X = Y \times Z$, so ist durch

$$\delta(gF,r) = r\,g\,e \in \mathbb{R}^n$$

ein Diffeomorphismus δ von X auf die Schale $S = \{x \in \mathbb{R}^n;\ 1 \leq |x| \leq 2\}$ definiert. δ induziert dann einen G-Isomorphismus der Algebra A_o/K auf eine invariante Unteralgebra der G-Algebra $C(X)$. Dabei ist klar, wie X und $C(X)$ als G-Objekte zu verstehen sind. Wir sehen, daß wir in beiden Fällen auf denselben Sachverhalt kommen:

Sei G eine lokal kompakte Gruppe, K eine kompakte Untergruppe und $Y = G/K$ der homogene Raum der Linksrestklassen und Z ein kompakter Raum.

Das cartesische Produkt $X = Y \times Z$ ist ein lokal kompakter G-Raum mit der Wirkung

$$g(y,z) = (gy,z),$$

falls $y = hK \in G/K$, $gy = ghK$. Für Funktionen f auf X definieren wir f^g durch $f^g(x) = f(gx)$ für $x \in X$, $g \in G$. Dann ist insbesondere $C_\infty(X)$ eine Banachsche G-Algebra. Wir betrachten nun eine G-invariante Banachsche Unteralgebra.

$$A \subset C_\infty(X)$$

für die wir stets die folgenden Eigenschaften voraussetzen (siehe [6], [7], [8]):

1) A ist eine involutive Banachsche Algebra (mit komplexer Konjugation als Involution) und es gilt $|a| = |\bar{a}| \geq |a|_\infty = \sup_x |a(x)|$ für alle $a \in A$.

2) Für jedes $a \in A$ ist die Abbildung $g \mapsto a^g$ stetig von G in A.

3) $|a^g| = |a|$ für alle $a \in A$, $g \in G$.

4) Das Ideal A_o der Funktionen aus A mit kompakten Trägern ist dicht in A.

5) Die Algebra B der von z aus Z unabhängigen Funktionen aus A ist - in natürlicher Weise als Teil von $C_\infty(Y)$ betrachtet - eine reguläre Funktionenalgebra auf Y, das Ideal B_o der kompakt getragenen Funktionen aus B ist dicht. Ebenso ist die Algebra C aller Funktionen w aus C(Z) mit $Aw \subset A$ eine reguläre Funktionenalgebra auf Z.

Die Algebra C in 5) ist offensichtlich genau die Unteralgebra der G-Invarianten in der adjungierten Algebra A^b. Ferner können wir das Tensorprodukt $B \otimes C$, also die von den Funktionen

$$b \otimes c : (y,z) \mapsto (b \otimes c)(y,z) = b(y)c(z),\ b \in B,\ c \in C,$$

erzeugte Algebra als Teil von A betrachten. Zu $(y,z) \in X$ und einer Umgebung W von (y,z) in X existiert dann ein $a = b \otimes c \in B \otimes C \subset A$ mit $a(y,z) = b(y)c(z) = 1$, supp $a \subset W$, insbesondere ist auch A eine reguläre Funktionenalgebra.

Daß die genannten Voraussetzungen im Fall der Heisenberggruppe erfüllt sind, kann man am einfachsten daraus erkennen, daß der Raum $C_o^\infty(V\times S)$ der glatten Funktionen mit kompakten Trägern im Streifen $V \times S$ im $\mathbb{R}^{n+1}$ für jedes Intervall $S = [a,b] \subset \mathbb{R}$ mit $o < a < 1 < 2 < b$ unter der Transformation $(y,z) \mapsto (zy,z)$ von $V \times S$ auf sich invariant ist, also ein dichtes Bild in der oben definierten Algebra $\tilde{A}$ hat. Es folgt ähnlich, daß etwa B hier alle Schwartzschen Funktionen auf V enthält und da C alle glatten Funktionen auf Z enthält.

Im Fall der Bewegungsgruppen M_n besteht C aus den Einschränkungen der radialen Funktionen aus A_o^G auf das Intervall $Z = [1,2]$. Für B_o kann man etwa die glatten Funktionen auf der Sphäre S^{n-1} nehmen, d.h. $B_o = C^\infty(S^{n-1})$. Ist b eine derartige Funktion und ist h eine glatte Funktion auf $\mathbb{R}^+$ mit kompakten Träger und $h(r) = 1$ für $1 \leq r \leq 2$, so ist die durch $f(x) = b(\frac{x}{|x|})h(|x|)$, $x \neq o$, $f(o) = 0$, definierte Funktion aus $A_o(\mathbb{R}^n)$

und definiert folglich in A eine Funktion $a = b \otimes 1 \varepsilon B$.

Im übrigen gilt für M_n folgendes: Ist H der Raum der hermiteschen Funktionen auf S^{n-1} und R die Algebra der glatten radialen Funktionen aus A_o, so ist $H \otimes R$ in A_o dicht. Zum Beweis betrachten wir den Raum D_o der glatten Funktionen auf dem $\mathbb{R}^n$ mit kompakten, zu $\{0\}$ disjunkten Trägern. D_o ist ein $SO(n,\mathbb{R})$ - invarianter dichter Teilraum von A_o und wird topologisch von seinen $SO(n,\mathbb{R})$ - irreduziblen endlichdimensionalen Unterräumen erzeugt. Sei $\{b_1,\ldots,b_m\}$ Basis eines solchen zur irreduziblen unitären Darsteillung $T : g \to T(g) = (t_{jk}(g))$, $1 \leq j,k \leq m$, gehörigen Unterraumes. Es gilt dann also

(2) $$b_j^g = \sum_{k=1}^{m} t_{jk}(g) b_k$$

für alle $g \varepsilon SO(n)$.

Sei wieder $e = \{1,o,\ldots,o\}$ der Einheitsvektor in x_1 - Richtung und $F \approx SO(n-1)$ die Fixgruppe von e. Wir wählen die Basis der b_j so, daß $t_{jk}(h) = \delta_{jk}$ ist für $h \varepsilon F$ und $1 \leq j,k \leq \ell$. Es ist dann auch $t_{jk}(gh) = t_{jk}(g)$ für $g \varepsilon SO(n)$, $h \varepsilon F$ für alle j und k, falls $k \leqq 1$.

Aus (2) folgt nun für alle $g \varepsilon SO(n)$, $h \varepsilon F$ und $r \varepsilon \mathbb{R}^+$:

$$b_j(g\cdot re) = b_j(gh\cdot re) = \sum_{k=1}^{m} t_{jk}(gh) b_k(re) =$$

$$= \sum_k \sum_i t_{ji}(g) t_{ik}(h) b_k(re).$$

Integrieren wir hier nach h über die Gruppe F, so erhalten wir

$$b_j(g\cdot re) = \sum_{k=1}^{\ell} t_{jk}(g) b_k(re).$$

Schreiben wir

$$\tau_{jk}(ge) = t_{jk}(g)$$

für $1 \leq j \leq m$, $1 \leq k \leq \ell$, so sind diese τ_{jk} offensichtlich aus H, ferner sind die durch

$$c_k(x) = b_k(|x|\ e)$$

aus R und es ist $b_j = \sum_{k=1}^{\ell} \tau_{jk} \otimes c_k$. Damit ist die Behauptung bewiesen.

Wir betrachten nun wieder allgemein die Gruppe G, den lokal kompakten G-Raum $X = Y \times Z$ mit $Y = G/K$ und die Unteralgebra A aus $C_\infty(X)$. Wir bilden dann die verallgemeinerte L^1-Algebra

$$L = L^1(G,A) \quad .$$

Wir bezeichnen für $z \in Z$ mit kz den Kern der Bahn $Y \times \{z\} \subset X$ in A, setzen also

$$kz = \{a \in A;\ a(y,z) = o,\ y \in Y\}.$$

Mit jz sei das kleinste abgeschlossene Ideal aus A bezeichnet, dessen Hülle gleich $Y \times \{z\}$ ist:

$$jz = \{a \in A;\ \text{supp}\, a \text{ kompakt und zu } Y \times \{z\} \text{ disjunkt}\}^-.$$

Wir setzen stets voraus, daß der Quotient kz/jz nilpotent ist, daß also ein $t \in \mathbb{N}$ existiert mit $(kz)^t \subset jz$.

Da kz und jz abgeschlossen und G-invariant sind, können wir die Algebren

$$L_{kz} = L^1(G, kz), \quad L_{jz} = L^1(G,jz)$$

bilden. Es gilt dann auch $L_{kz} \subset L_{jz}$.

(3) <u>Ist</u> E <u>ein topologisch irreduzibler</u> L-<u>Modul, so existiert ein</u> $z \in Z$ <u>mit</u> $L_{kz}E = o$.

Beweis: Die Unteralgebra C aus A^b läßt sich aufgrund der Einbettung $A^b \to L^b$ auch als zentrale Unteralgebra von L^b betrachten. Somit ist der in E dichte Untermodul $LE = E_o$ ein C-Modul. Aus der topologischen Irreduzibilität von E folgt dann für jedes $f \in C$ entweder $fE_o = o$ oder $\overline{fE_o} = E$. Da C eine reguläre Funktionalalgebra auf Z ist, existiert somit ein $z \in Z$, so daß E_o vom Ideal C_z aller $f \in C$ mit $z \notin \operatorname{supp} f$ annulliert wird. Wegen $L_{jz} \subset LC_z^{\#}$ folgt $L_{jz}\, E = o$, also $L_{kz}^t\, E = o$ und folglich $L_{kz}\, E = o$.

Folgerung:

(4) <u>Jedes abgeschlossene maximale Linksideal aus</u> L <u>enthält ein Ideal</u> L_{kz}. <u>Ist</u> $K = \{e\}$ <u>oder</u> G <u>kompakt, so ist ein abgeschlossenes Linksideal</u> Λ <u>aus</u> L <u>dann und nur dann in einem abgeschlossenen maximalen Linksideal enthalten, wenn ein</u> $z \in Z$ <u>existiert mit</u>

$$(\Lambda + L_{kz})^- \neq L.$$

Beweis: Ist Λ maximal, so ist $E = L/\Lambda$ ein topologisch irreduzibler L-Modul, also $L_{kz}E = o$, d.h. $L_{kz}L \subset \Lambda$ und damit $L_{kz} \subset \Lambda$ für ein $z \in Z$.

Ist umgekehrt L_{kz} in Λ enthalten, so können wir im Quotienten L/L_{kz} rechnen. Nun ist

$$L/L_{kz} \cong L^1(G, A/kz)$$

und $A/kz = A_z \subset C_\infty(Y)$, mit einer Unteralgebra A_z, wie sie in [7], Teil III, betrachtet wurde. Es folgt nun leicht aus den Ergebnissen in [7], daß jedes abgeschlossene Linksideal in $L^1(G, A_z)$ in maximalen enthalten ist.

Die Unteralgebra B der von $z \in Z$ unabhängigen Funktionen aus A läßt sich auch als G-invariante Unteralgebra von $C_\infty(Y)$ betrachten. Die Algebra $L_B = L^1(G,B)$ ist dann in L enthalten und hat genau die Struktur der in [7], Teil III, untersuchten Klasse von Algebren. Man kann diesen Umstand benutzen, um den Verband der Linksideale von L mit ähnlichen Methoden wie in [7] zu studieren. Der Einfachheit halber wollen wir das hier nur für den Fall skizzieren, daß G unimodular und die kompakte Untergruppe K trivial, d.h. gleich $\{e\}$ ist. Das ist z.B. der Fall für die Heisenberggruppen, sowie für $G = M_2$. Die Algebra L_B ist dann einfach und symmetrisch und wird von ihren minimalen Idempotenten erzeugt.

Sei nun also $X = G \times Z$. Wie in [7] setzen wir für $a,b \in A$

$$(a \circ b)(g) = a^g\bar{b} \in A,$$

so daß $g \mapsto (a \circ b)(g)$ eine stetige Funktion $a \circ b$ von G in A definiert. Es ist

$$(a \circ b)(g)(y,z) = a(gy,z)\overline{b(y,z)}.$$

Nun wählen wir weiter ein festes u aus B_o mit $u \geq 0$ und $\int_G u(g)^2 dg = 1$. Dann ist $p = u \circ u$ aus L und

$$(p * p)(g) = \int_G u^g u^{t^{-1}} u^{t^{-1}} u\, dt = u^g u$$

d.h. $p * p = p$, $p^*(g) = \{(u^{g^{-1}} u)^-\}^g = p(g)$, somit ist p ein hermitesches Idempotent aus L. Für $f \in L$ ist

$$(p * f)(g) = \int u^g u^t f(t)\, dt = u^g i(f)$$

mit $i(f) = \int (u^{\#} \bar{f})(t)\, dt \in A$. Es ist also $p * f = u \circ i(f)$, d.h. $p * L$ besteht aus Elementen der Form $u \circ a$ mit gewissen $a \in A$. Wir definieren

$$A_1 = \{a \in A;\ u \circ a \in L\}.$$

Offensichtlich ist A_1 ein linearer Unterraum von A und es gilt

$$p * L = u \circ A_1 = \{u \circ a;\ a \in A_1\}.$$

Wie in [7] können wir auch hier auf A_1 eine Norm $\|a\|$ definieren, vermöge

$$||a|| = |u \circ a|_1 = \int_G |u^g\, \bar{a}|\,dg.$$

Damit ist denn A_1 ein zu $p * L$ isomorpher Banachscher Raum. Darüber hinaus gilt

(5) a) $A_1 \subset \{a \in A;\ \int |a^g\, \bar{a}|\,dg < \infty\}$.

b) Für a und b aus A_1 ist $a \circ b \in L$.

c) A_1 ist ein G-invariantes Ideal in A.

Beweis: Für $a \in A_1$ sind $u \circ a$ und $a \circ u = (u \circ a)^*$ aus L, also auch $(a \circ u) * (u \circ a)$. Sind nun $a \circ b$ und $c \circ d$ für $a,b,c,d \in A$ aus L, so gilt allgemein für $g \in G$, $y \in Y$, $z \in Z$:

$$\begin{aligned}(a \circ b) * (c \circ d)(g)(y,z) &= \int a(gy,z)\overline{b(t^{-1}y,z)}c(t^{-1}y,z)\overline{d(y,z)}\,dt \\ &= a(gy,z)\overline{d(y,z)} \int_G \overline{b(t,z)}c(t,z)\,dt \\ &= (a \circ d)(g)(y,z)(c|b)(z)\end{aligned}$$

d.h.

(6) $(a \circ b) * (c \circ d) = (c|b)(a \circ d)$

mit der Funktion

$$(c|b) : z \mapsto \int_G c(t,z)\overline{b(t,z)}\,dt.$$

Aus der Voraussetzung über u folgt $(u|u) = 1$, folglich $(a \circ u) * (u \circ a) = a \circ a \in L$, also $\int a^g \bar{a}|\,dg < \infty$. Ebenso $(a \circ u) * (u \circ b) = a \circ b \in L$ für $a,b \in A_1$. Für $a \in A_1$ und $f \in A$ folgt

$$|u \circ (af)|_1 = \int |u^g\, \overline{af}|\,dg \leqq \int |u^g\, \bar{a}|.|f|\,dg = |u \circ a|_1 |f|$$

also $af \in A_1$ und $||af|| \leqq ||a|| \cdot |f|$. Schließlich folgt für $a \in A_1$ und $h \in G$:

$$||a^h|| = \int |u^g\, \bar{a}^h|\,dg = \int |u^{gh^{-1}}\, \bar{a}|\,dg = \int |u^g\, \bar{a}|\,dg = ||a||.$$

Damit ist (5) bewiesen.

Als nächstes betrachten wir die abgeschlossene *-Unteralgebra

$$P = p * L * p.$$

Sie besteht aus allen Elementen der Form

$$(u \circ a) * (u \circ u) = (u \mid a) u \circ u$$

d.h. es ist

$$P = Wp = \{wp;\ w \in W\}$$

mit der Algebra W aller $w \in C(Z)$ der Form $w = (u|a)$, d.h. $w(z) = \int_G u(g)\,\overline{a(g,z)}\,dg$ mit Elementen $a \in A_1$. Es ist leicht zu sehen, daß W die Algebra C enthält. Darüber hinaus gilt:

(7) A_1 <u>ist ein W-Modul und mit</u> $||w|| = |wp|_1$ <u>für</u> $w \in W$ <u>gilt</u> $||wa|| \leq ||w|| \cdot ||a||$ <u>für alle</u> $w \in W$ <u>und</u> $a \in A_1$.

Das folgt aus der Identität $(wp)*(u \circ a) = u \circ (wa)$.

Wie in |7| betrachten wir nun ein abgeschlossenes Linksideal $\Lambda \subset L$ und definieren den Unterraum $\beta(\Lambda) \subset A_1$ durch

$$p * \Lambda = u \circ \beta(\Lambda),$$

d.h. $\beta(\Lambda)$ besteht aus allen $a \in A_1$ mit $u \circ a \in \Lambda$. Aus

$$u \circ (W\beta(\Lambda)) = Wu \circ \beta(\Lambda) = Pp * \Lambda \subset p * \Lambda = u \circ \beta(\Lambda)$$

folgt $W\beta(\Lambda) \subset \beta(\Lambda)$, bzw. $W\beta(\Lambda) = \beta(\Lambda)$, da $1 \in W$.
Somit ist $\beta(\Lambda)$ ein W-Modul in A_1.
Ist umgekehrt V ein W-Untermodul von A_1, so ist

$$\gamma(V) = L*(u \circ V)^-$$

ein abgeschlossenes Linksideal in L. Aus

$$u \circ \beta\gamma(V) = p * L * (u \circ V)^{-} = W u \circ V = u \circ WV = u \circ V$$

folgt $\beta\gamma(V) = V$. Umgekehrt folgt für ein abgeschlossenes Linksideal Λ aus L:

$$\gamma\beta(\Lambda) = L * (u \circ \beta(\Lambda))^{-} = L * p * \Lambda = (L\Lambda)^{-},$$

da $L * p * L$ in L dicht ist. Es folgt:

(8) Die Abbildungen γ and β stellen eine Bijektion her zwischen den Verbänden der abgeschlossenen W-Untermoduln von A_1 und der abgeschlossenen Linksideale Λ mit $\Lambda = (L\Lambda)^{-}$.

Besitzt L also eine approximierende Eins, so sind die Verbände der abgeschlossenen Linksideale in L und der abgeschlossenen W-Untermoduln von A_1 isomorph, insbesondere gilt dann:

(9) L enthält dann und nur dann ein abgeschlossenes Linksideal ohne maximale abgeschlossene Ober-Linksideale, wenn A_1 einen echten abgeschlossenen W-Untermodul D enthält, so daß $D + k_1 z$ für jedes $z \in Z$ in A_1 dicht ist. Hier ist

$$k_1 z = A_1 \cap kz.$$

Das folgt aus (8) und (4) wegen $\beta(L_{kz}) = k_1 z$.

Ist G kompakt, so können wir z.B. $u \equiv 1$ wählen. Auf jeden Fall ist dann $A_1 = A$, B die Algebra der nur von $y \in Y = G$ abhängigen Funktionen und $C = W$ die Algebra der nur von $z \in Z$ abhängigen Funktionen aus A. Im Fall der Bewegungsgruppe M_2 erhält man aus (9) insbesondere das Theorem 3 aus [9]. Das Hauptergebnis in [9] ist natürlich der Nachweis der Existenz eines Untermodul D wie in (9), d.h. für die M_2 der Existenz eines abgeschlossenen echten A_0^G-invarianten Untermoduls D von $A_0(\mathbb{R}^2) = A_0$, für den $D + k(S_r)$ für jedes $r > 0$ in A_0 dicht ist, $k(S_r)$ der Kern des

Kreises S_r vom Radius $r > o$ in A_o.

Im Fall der Heisenberggruppe H_1 ist $L = L^1(\mathbb{R},A)$ ein Quotient von $L^1(H_1)$, in dem die $\mathbb{R}$-Algebra A wie folgt definiert war: Mit $Z = [1,2]$, $Y = \mathbb{R}$ ist $X = \mathbb{R} \times Z$ mit der Wirkung $t(y,z) = (y+t,z)$ von $\mathbb{R}$ auf X. Ist $k = k(X)$ der Kern von X in der Fourier-Algebra $A(\mathbb{R}^2)$, und wirkt $\mathbb{R}$ auf $A(\mathbb{R}^2)$ vermöge $f^t(y,z) = f(y+zt,z)$, so erhält man einen $\mathbb{R}$-Isomorphismus ρ der $\mathbb{R}$-Algebra $A(\mathbb{R}^2)/k$ auf A in folgender Weise: Ist $a \in A(\mathbb{R}^2)$ und $a_k = a+k$ das Bild von a in $A(\mathbb{R}^2)/k$, so ist $\rho(a_k) \in A$ durch

$$\rho(a_k)(y,z) = a(zy, z)$$

definiert. Dann ist z.B.

$$\rho(a_k^t)(y,z) = \rho(a_k^t)(y,z) = a^t(zy, z) = a(zy+zt, z) =$$

$$= a(z(y+t),z) = \rho(a_k)(y+t, z) = \rho(a_k)^t(y,z).$$

Der Algebra B der von $z \in Z$ unabhängigen Funktionen entspricht in $A(\mathbb{R}^2)/k$ die Algebra der Restklassen von Funktionen $b \in A_+ \subset A_o(\mathbb{R}^2)$ mit $b(y,z) = b(z^{-1}y, z)$ für $z \in Z$. Ist φ eine Schwartz'sche Funktion auf $\mathbb{R}$ und ist q eine glatte Funktion auf $\mathbb{R}$ mit kompaktem Träger in $\mathbb{R}^+$ und $q(x) = 1$ in einer Umgebung von Z, so ist durch $b(y,z) = \varphi(z^{-1}y)q(z)$ für $z > o$ eine Funktion b mit $\varphi = \rho(b) \in B$ definiert, es ist also $S(\mathbb{R}) \subset B$. Ist $\varphi \in S(\mathbb{R})$ positiv mit kompaktem Träger und $\int \varphi(t)^2 dt = 1$, so genügt die hierdurch definierte Funktion u aus B den oben an die ebenfalls mit u bezeichnete, zur Definition von $p = u \circ u$ gewählten Funktion gestellten Bedingungen.

Für $a \in A(\mathbb{R}^2)$ ist dann

$$(u|\rho(\bar{a}_k))(z) = \int \varphi(t)\ a(zt,z)dt = \int \varphi(z^{-1}t)\ a(t,z)dt = \int f(t,z)dt$$

mit der durch $f(t,z) = \varphi(z^{-1}t)\ a(t,z)\ q(z)$ für $z > o$ definierten Funktion. Offensichtlich ist f aus $A_+(\mathbb{R}^2)$ und hat einen kompakten Träger.

Es folgt, daß die durch $F(z) = \int f(t,z)dt$ definierte Funktion aus $A(\mathbb{R})$ ist, mit kompaktem Träger in $\mathbb{R}^+$. Folglich liegt $(u|\rho(\bar{a}_k))$ in $A(\mathbb{R})/k(Z)$, d.h. es ist

$$W = C = A(\mathbb{R})/k(Z).$$

Definieren wir $U \in A(\mathbb{R}^2)$ durch (s.o.)

$$U(y,z) = \varphi(z^{-1}y)q(z),$$

so hat U kompakten Träger und die Algebra A_1 ist unter ρ das Bild der Algebra $A(\mathbb{R}^2)_1/k$ mit

$$A(\mathbb{R}^2)_1 = \{a \in A(\mathbb{R}^2);\ \int |(U^t\ a)_k| dt < \infty\}.$$

Betrachten wir statt der Quotienten von $L^1(H_1)$ zweiseitige Ideale der Form $L^1(\mathbb{R},k(X))$ für $\mathbb{R}$-invariante Teile $X = \mathbb{R} \times T$ mit $T = \mathbb{R} \setminus]a,b[$ für offene Intervalle $]a,b[\subset \mathbb{R}$, so führt die Frage nach den Linksidealen ganz analog auf folgendes Problem: Man betrachte die Fourier-Algebra $A(\mathbb{R}^2) = A(\mathbb{R}) \hat{\otimes} A(\mathbb{R})$ als $A(\mathbb{R})$-Modul vermöge $(au)(y,z) = a(z)\ u(y,z)$ für $u \in A(\mathbb{R}^2)$ und $a \in A(\mathbb{R})$. Gegeben sei ein $A(\mathbb{R})$-Untermodul $A_1 \subset A(\mathbb{R}^2)$, der gewissen Voraussetzungen genügt, etwa ein Ideal in $A(\mathbb{R}^2)$ ist oder sämtliche Funktionen mit kompakten Trägern enthält. Ferner habe A_1 eine Norm, unter der A_1 ein Banachscher $A(\mathbb{R})$-Modul ist. Unter welchen Bedingungen existieren dann abgeschlossene $A(\mathbb{R})$-Untermoduln in A_1, die keine maximalen Obermoduln besitzen ? Dieses Problem ist im Fall $A_1 = A(\mathbb{R}^2)$ von Dixmier in [1] untersucht. Dort handelt es sich übrigens "in Wirklichkeit" ebenfalls um Darstellungen der Heisenberggruppe: Wie in [3] beschrieben, lassen sich die stetigen irreduziblen Darstellungen von $L^1(H_1)$ in Banachschen Räumen über die primitiven Quotienten von $L^1(H_1)$ faktorisieren. Diese sind kanonisch isomorph zu verallgemeinerten Algebren [4] $L^1(\mathbb{R}^2, \mathbb{C}; p)$ mit den Faktorsystemen $p_{x,y} = e^{iz_1 x_1 y_2}$ für $x = (x_1,x_2)$, $y = (y_1,y_2)$ $\mathbb{R}^2$, z fest, reell. Alle diese $L^1(\mathbb{R}^2, \mathbb{C}; p)$

sind zu $\Gamma(\mathbb{R})$ isomorph und besitzen genau eine irreduzible unitäre Darstellung, die sich in $L^2(\mathbb{R})$ durch

$$\rho(f)\,\xi(t) = \iint f(x,y)e^{i\,yt}\,\xi(x+t)\,dx\,dy$$

realisieren läßt. Dieselbe Formel gibt eine Darstellung in $L^p(\mathbb{R})$ für jedes p, mit $1 \leq p \leq \infty$. Wählt man nun ein $\xi \in L^p(\mathbb{R})$ mit $\xi \notin L^\infty(\mathbb{R})$, so ist

$$M = \{f \in L^1(\mathbb{R}^2) = L^1(\mathbb{R}^2, \mathbb{C};\, p);\ \rho(f)\xi = o\}$$

ein abgeschlossenes Linksideal in $L^1(\mathbb{R}^2, \mathbb{C};\, p)$, das sich auch als abgeschlossener Unterraum der abelschen Algebra $L^1(\mathbb{R}^2)$ betrachten läßt, der zweiten Variablen translationsinvariant ist. $\hat{M} = \{\hat{f};\ f \in M\}$ ist dann ein abgeschlossener $A(\mathbb{R})$-Untermodul von $A(\mathbb{R}^2)$, der keine maximalen Obermoduln besitzt.

II.

Wir betrachten jetzt die einfachen symmetrischen Algebren $L^1(G,Q)$, in denen Q eine reguläre Banachsche Unteralgebra von $C_\infty(G)$ ist, in der das Ideal Q_o der kompakt getragenen Funktionen dicht ist. Diese Algebren und ihre abgeschlossenen Linksideale sind in [7] genau beschrieben, insbesondere wird in [7], Teil I, gezeigt, daß Q eine Segal-Algebra Q_1 enthält, so daß ein Element f aus

$$\mathcal{L} = L^1(G,Q)$$

genau dann den Rang 1 hat (d.h. der Operator $\rho(f)$, Bild von f bei der kanonischen regulären Darstellung von $\mathcal{L}$ in $L^2(G)$, hat den Rang 1), wenn $f = a \circ b$ ist mit $a, b \in Q_1$, $a \neq o$, $b \neq o$. Hier ist $a \circ b$ die durch $(a \circ b)(x) = a^x\,\bar{b}$ definierte stetige Abbildung von G in Q. Das Ideal $\mathcal{E}$ der Operatoren endlichen Ranges besteht aus allen Linearkombinationen solcher $a \circ b$ und $\mathcal{E}$ ist dicht in $\mathcal{L}$.

Für jedes $f \in L$ ist $\rho(f)$ ein kompakter Operator, insbesondere ist $\rho(f) = \sum_{j=1}^{\infty} \lambda_j E_j$ mit reeller Nullfolge $\{\lambda_j\}$ und paarweise orthogonalen Projektoren endlichen Ranges E_j, falls f hermitesch ist. Ein orthogonaler Projektor E vom Rang 1 in $L^2 = L^2(G)$ ist stets ein zerfallender Integraloperator der Form

$$E\,\xi(x) = \int \alpha(x)\overline{\alpha(y)}\,\xi(y)\,dy = (\xi|\alpha)\alpha(x)$$

mit $\alpha \in L^2(G)$, $|\alpha|_2 = 1$. Schreiben wir hierfür E_α, so ist insbesondere $\rho(a \circ a) = E_a$ für $a \in Q$, mit $|a|_2 = 1$. Die Eigenprojektoren E_j sind dann endliche Summen solcher E_α, somit können wir $\rho(f)$ auch als $\Sigma \lambda_j E_{\alpha_j}$ schreiben mit einer orthonormierten Familie $\{\alpha_j\} \subset L^2$. Wir bemerken, daß Q_1 ein dichter Teilraum von L^2 ist. Wählen wir eine Familie $\{a_j\} \subset Q_1$ mit $(a_j|a_k) = \delta_{jk}$, so ist klar, daß $f = \Sigma \mu_j a_j \circ a_j$ in L liegt, falls $\{\mu_j\}$ schnell genug gegen Null konvergiert, falls also z.B. $\Sigma|\mu_j||a_j \circ a_j| < \infty$ ist. In diesem Fall ist $\rho(f) = \Sigma \mu_j E_{a_j}$. Wir werden nun zeigen, daß für <u>jedes</u> hermitesche f aus L die in der Eigenentwicklung $\rho(f) = \Sigma \lambda_j E_{\alpha_j}$ auftretenden α_j in Q_1 liegen.

<u>Satz</u>. <u>Ist f ein hermitesches Element aus L, so ist das Spektrum von f eine reelle Nullfolge $\{\lambda_j\}_{j\in\mathbb{N}}$. Zu jedem λ_j des Spektrums existieren endlich viele Elemente $v_{jk} \in Q_1$, so daß die Familie $\{v_{jk}\}$ aller dieser v_{jk} in $L^2(G)$ orthonormiert ist und</u>

$$\rho(f) = \sum_{j,k} \lambda_j\, \rho(v_{jk} \circ v_{jk})$$

<u>gilt</u>.

Beweis: Da L symmetrisch ist haben f und $\rho(f)$ dasselbe Spektrum, da $\rho(f)$ kompakt und hermitesch ist, handelt es sich dabei also um eine reelle Nullfolge $\lambda = \{\lambda_j\}$. Die Resolvente $R(z)$ von f ist im Komplement von λ in $\mathbb{C}$ holomorph. Sei K ein Kreis mit dem Mittelpunkt $\lambda_j \neq 0$, der keine weiteren λ_k enthält. Dann ist

$$f_j = (2\pi i \lambda_j)^{-1} \int_K f R(\zeta) d\zeta \in L$$

und falls $\rho(f) = \sum_1^\infty \lambda_n E_n$ mit Projektoren E_j:

$$\rho(f_j) = \Sigma (2\pi i \lambda_j)^{-1} \int_K \frac{\lambda_n}{\zeta - \lambda_n} d\zeta E_n = E_j .$$

Folglich ist f_j ein hermitesches Idempotent endlichen Ranges aus L, also $f_j L f_j$ eine einfache Algebra endlicher Dimension mit f_j als Einselement. Aus [7], Teil I, folgt also, daß $f_j = \sum_{k=1}^m v_{jk} \circ v_{jk}$ ist mit $v_{jk} \in Q_1, (v_{jk}|v_{j1}) = \delta_{k1}$. Damit ist der Satz bewiesen.

Lassen wir für die λ_j Vielfachheiten zu, so könen wir etwas einfacher schreiben

$$\rho(f) = \sum_j \lambda_j \rho(v_j \circ v_j).$$

Da f durch die λ_j und v_j eindeutig bestimmt ist, erhalten wir damit auch eine Art Eigenfunktionen-Entwicklung für f:

$$(10) \qquad f \sim \sum_j \lambda_j (v_j \circ v_j).$$

In der Regel wird natürlich die Reihe auf der rechten Seite in L nicht gegen f konvergieren. Nehmen wir der Einfachheit halber an, daß alle λ_j verschieden sind, so folgt aus der Symmetrie von L, daß die von f erzeugte abgeschlossene Unteralgebra <f> alle $v_j \circ v_j$ enthält, also auch die von diesen erzeugte abgeschlossene Unteralgebra. Dagegen steht man vor folgendem

Problem: Liegt die hermitesche Funktion f mit der Eigenfunktionen-Entwicklung (10) in der von allen Eigenfunktionen $v_j \circ v_j$ erzeugten abgeschlossenen Unteralgebra F von L ?

Für solche Funktionen, die wir sanft nennen wollen, haben wir folgendes Resultat:

(11) Sei $P(t)$ ein von Null verschiedenes Polynom ohne konstantes Glied und sei f eine sanfte Funktion und $P(\lambda_j) \neq 0$ für alle $\lambda_j \neq 0$ in der Entwicklung $f \sim \Sigma\, \lambda_j(v_j \circ v_j)$. Dann liegt f in dem von $P(f)$ erzeugten abgeschlossenen Linksideal Θ.

Beweis: Mit $f_j = v_j \circ v_j$ ist $P(f) \sim \Sigma\, P(\lambda_j) f_j$, also liegt $P(\lambda_j) f_j = f_j\, P(f)$ in Θ und somit auch f_j. Es folgt $f \in F \subset \Theta$. Insbesondere gilt also:

(12) Liegt eine positive Potenz einer sanften Funktion f im abgeschlossenen Linksideal Λ, so auch f.

Wenn man beweisen könnte, daß hermitesche Elemente aus L stets sanft sind, so würde man aus (12) als unmittelbare Folgerung das Lemma (5.13) aus [2] und seine Folgerungen erhalten:

(13) Sei $\{q_t\}_{t>o}$ eine Faltungshalbgruppe in L aus sanften hermiteschen Elementen q_t. Ist $\lim_{t \to o} f\, q_t = f$ für jedes $f \in L$, so ist $Lq_t = L$ für jedes $t > o$.

Beweis: Sei $Lq_s = \Lambda$ für ein festes $s > o$. Ist $t > o$ und $n\,t = s + r > s$ für $n \in \mathbb{N}$, so folgt $q_t^n = q_{nt} = q_r\, q_s \in \Lambda$, nach (12) also $q_t \in \Lambda$ und folglich $f\, q_t \in \Lambda$ für jedes $f \in L$ und alle $t > o$. Somit folgt auch $f \in \Lambda$, d.h. $\Lambda = L$

References

[1] J. Dixmier: Remarques sur un theorème de Wiener. Bull. Sc. Math. France 84 (1960) 35-40.

[2] A. Hulanicki: Subalgebra of $L_1(G)$ associated with Laplacian on a Lie group. Coll. Math. 31 (1974) 259-287.

[3] H. Leptin: On group algebras of nilpotent groups. Studia Math. 47 (1973) 37-49.

[4] H. Leptin: Verallgemeinerte L^1-Algebren und projektive Darstellungen lokal kompakter Gruppen. Inventiones Math. 3 (1967), 257-281, 4 (1967), 68-86.

[5] H. Leptin: Symmetrie in Banachschen Algebren. Archiv Math. 27 (1976) 394-400.

[6] H. Leptin: Ideal theory in group algebras of locally compact groups. Inventiones Math. 31 (1976) 259-278.

[7] H. Leptin: On onesided harmonic analysis in non commutative locally compact groups. Erscheint im J. Reine u. Angew. Math.

[8] H. Leptin, D. Poguntke: Symmetry and nonsymmetry for locally compact groups. Erscheint im J. Funct. Anal.

[9] Y. Weit: On the one sided Wiener's theorem for the motion groups. Preprint (1978).

C^∞ PARAMETRIX ON LIE GROUPS AND TWO STEPS FACTORIZATION ON CONVOLUTION ALGEBRAS

by Paul MALLIAVIN

Institut Henri Poincaré

11, rue Pierre et Marie Curie

75005 Paris

We denote by G a connected real Lie group. We denote by $\mathcal{D}(G)$ the convolution algebra of smooth functions of compact support. We say that <u>factorization at</u> step k holds true for $\mathcal{D}(G)$ if $\forall \varphi \in \mathcal{D}(G)$. we can find $\Psi_s, \Theta_s \in \mathcal{D}(G)$, $1 \le s \le k$, such that

$$\varphi = \sum_{s=1}^{k} \Psi_s * \Theta_s$$

In [1] it was proved that factorization at step $2^{\dim(G)}$ is true. Using a parametrix for an infinite order P.D.E. in $\mathbb{R}^n$, it was also proved that on $\mathbb{R}^n$ factorization at step 2 is true. The construction of the parametrix was made by the Fourier transform. We shall replace here the Fourier transform by a symbolic calculus on an heat semi group in the spirit of [2], [4], [5], and the same qualitative properties as in the $\mathbb{R}^n$ case of the parametrix will be proven using the same P.D.E. as in $\mathbb{R}^n$. I am indebted to Jacques Dixmier for helpful conversation during the writing of this paper. See also [8] for related results.

1. <u>Notations</u>

We choose a basis $A_1 \ldots A_m$ of the right invariant vector fields on G, and we consider the elliptic operator Δ defined in the left enveloping algebra by

$$\Delta = \frac{1}{2} \sum_{k=1}^{m} A_k^2 .$$

Using on G a left invariant Haar measure dg, then Δ is formally symmetric on $L^2(G)$. We denote by $p_t(g)$ the elementary solution of the heat kernel defined by

$$\frac{\partial p_t}{\partial t} = \Delta p_t \qquad t > 0$$

$$p_t(g)\,dg \to \delta_e \quad \text{when} \quad t \to 0 \quad \text{(e the identity)}.$$

As in [1] we shall consider an infinite subsequence Λ of the sequence $\{2^n\}_{n=1}^{+\infty}$; associate to Λ the entire function

1.1 $$f_\Lambda(\zeta) = \prod_{\lambda\in\Lambda} (1+\frac{\zeta}{\lambda}) \qquad \zeta \in \mathbb{C}\ .$$
$$= \sum_{n\geq o} a_{n,\Lambda}\, \zeta^n\ .$$

We associate also to Λ the infinite order P.D.E., defined by

1.2 $$L_\Lambda = f_\Lambda(-\Delta) = \sum_{n\geq o} a_n(-\Delta)^n\ .$$

We finally introduce the inverse Laplace transform k_Λ of f_Λ^{-1} .

1.3 $$k_\Lambda(t) = \int_{-\infty i}^{+\infty i} \frac{1}{f(\zeta)}\, e^{\zeta t}\, \frac{d\zeta}{2i\pi} \qquad t>0,\quad \mathrm{Re}\zeta > 0,$$

and we define

1.4 $$q_\Lambda(g) = \int_0^{+\infty} p_t(g)\, k_\Lambda(t)\,dt$$

2. <u>Theorem</u>: <u>The integrals</u> 1.3, 1.4 <u>are absolutely convergent</u>,

2.1 $$||q_\Lambda||_{L^1(G)} = 1$$

2.2 q_Λ <u>is</u> C^∞ <u>on</u> G.

2.3 <u>Let</u> A <u>be a compact on</u> G, $e \notin A$ <u>then</u> q_Λ <u>belongs to the Gevrey class</u> 2 <u>on</u> A, <u>uniformly in</u> Λ.

2.4 <u>Corollary: Factorization at step two holds true in</u> (G).

3. <u>Symbolic calculus on a semi-group</u>

We shall use the symbolic calculus on the semi group of convolution on G defined by p_t: we have in fact

3.1 $p_t * p_{t'} = p_{t+t'}$

We shall denote by M the C^∞ functions u defined on $\mathbb{R}^+$, such that

$$u^{(k)}(t) \to 0 \quad \text{when } t \to 0 \quad k = 0,1,\ldots$$

$$||u^{(k)}||_{L^1(R)} < +\infty.$$

We associate to $u \in M$ the kernel

$$p_u(g) = \int_0^{+\infty} p_t(g)\; u(t)\; dt.$$

Then 3.1 implies that

3.2 $p_u * p_v = p_{u*v}$

In this identity the convolution in the left hand side is taken on G, in the right hand side on R^+. We have also

3.3 <u>Lemma</u>: $(-\Delta) p_u = p_{u'}$

Proof:

We introduce

$$u_\varepsilon(t) = u(t-\varepsilon) \quad \text{if} \quad t > \varepsilon$$
$$= 0 \quad \text{if} \quad t < \varepsilon .$$

Then by the identity

$$\frac{\partial p_t}{\partial t} = \Delta p_t$$

and by integration by part

$$-\Delta p_{u_\varepsilon} = p_{u'_\varepsilon}$$

Now using elliptic estimates we have when $t \to 0$

$$||\Delta p_t||_{L^\infty(G)} = 0(t^{-n}) \qquad (n = \dim G)$$

As $u(t) = 0(t^k)$ for all k we have

Δp_{u_ε} converges uniformly to Δp_u which proves 3.3.

3.4 Lemma: If $u \in M$, then $p_u \in C^\infty(G)$

Proof:

$$||\Delta^k\, p_u||_{L^1(G)} \leq ||u^{(k)}||_{L^1(R^+)} < +\infty$$

Then the Sobolev's immersion theorem gives the result.

4. Properties of k

4.1 Lemma: $k_\Lambda \in M$

Proof: $\int_0^{+\infty} |k^{(s)}(t)|\, dt \leq \sup_t (1 + t^2) \left| k_\Lambda^{(s)}(t) \right|$

$$(1+t^2)k_\Lambda^{(s)}(t) = \int_{-i\infty}^{+i\infty} \left[\frac{\zeta^s}{f_\Lambda(\zeta)} - \left(\frac{\zeta^s}{f_\Lambda(\zeta)}\right)''\right] e^{\zeta t} \frac{d\zeta}{2i\pi}$$

We shall integrate on $\mathrm{Re}\zeta = 0$, then the fact that for all n

$$\frac{\zeta^n}{f_\Lambda(\zeta)} \to 0 \quad \text{for} \quad |\zeta| \to +\infty, \quad \frac{-1}{2} < \mathrm{Re}\zeta < \frac{1}{2}$$

combined with the Cauchy's theorem for the evaluation of the second derivative proves the statement.

When $t \to 0$ we integrate on $\mathrm{Re}\zeta = \frac{1}{t}$ and we get the majoration

$$|k_\Lambda^{(s)}(t)| \leq \int_{\mathrm{Re}\zeta = \frac{1}{t}} \left|\frac{\zeta^s}{f(\zeta)}\right| |d\zeta| = o(1)$$

4.2 <u>Lemma</u>

$$||k_\Lambda||_{L^1} = 1$$

<u>Proof</u>:

We remark that $f_\Lambda^{-1}(\zeta)$ has a natural form as a product, therefore k_Λ can be expressed as a convolution product

$$k_\Lambda = \prod_{\lambda\varepsilon\Lambda}{}^{*} h_\lambda$$

where

$$h_\lambda(t) = \int_{-i\infty}^{+i\infty} \frac{e^{\zeta t}}{1+\zeta/\lambda} \frac{d\zeta}{2i\pi} = \lambda e^{-\lambda t} .$$

We have

$$||h_\lambda||_{L^1} = 1, \quad \text{which proves 4.2}$$

5. <u>Fundamental solution of</u> L_Λ.

Denote

$$f_\Lambda^s(\zeta) = \prod_{2<\lambda<2^s} \left(1 + \frac{\zeta}{\lambda}\right)$$

Denote by M_n a logarithmic convex sequence of positive number such that

5.0 $\quad \Sigma\, a_{n,\Lambda} R^n M_n < +\infty \quad$ for all $R < +\infty$,

and by $\mathcal{D}_{M_n}$

$$\mathcal{D}_{M_n} = \{\varphi \varepsilon\, \mathcal{D}(G);\ \ ||\Delta^n\varphi||_{L^\infty} \leq c\, M_n\}$$

Define

$$L_\Lambda^s = f_\Lambda^s(-\Delta).$$

then, where $s \to +\infty$,

$$||L_\Lambda^s\varphi - L_\Lambda\varphi||_{L^\infty} \to 0 \qquad \text{for all } \varphi\varepsilon\, \mathcal{D}_{M_n}$$

5.1 <u>Proposition</u>

$$\langle L_\Lambda\varphi,\ q_\Lambda\rangle = \varphi(e) \qquad \text{for all } \varphi\varepsilon\, \mathcal{D}_{M_n}$$

<u>Proof</u>:

$$\langle L_\Lambda\varphi,\ q_\Lambda\rangle = \int (L_\Lambda\varphi)(g)\ q_\Lambda(g)\ dg$$

$$= \lim_{s\to\infty} \int (L_\Lambda^s\ )(g) q_\Lambda(g)\ dg$$

$$I_s = \int (L_\Lambda^s\varphi)(g) q_\Lambda(g) dg = \int_0^{+\infty} k_\Lambda(t)\ dt \langle L_\Lambda^s\varphi,\ p_t\rangle$$

Using the fact that Δ is symmetric

$$I_s = \int_0^{\infty} k_\Lambda(t)\,dt < \varphi,\, L_\Lambda^s\, p >$$

$$I_s = \int_0^{\infty} k_\Lambda(t)\,dt < \varphi,\, f_\Lambda^s(-\tfrac{d}{dt}) p_t$$

$$\int_0^{\infty} k_\Lambda(t)\; f_\Lambda^s\; (-\tfrac{d}{dt}) < \varphi,\, p_t > dt\ .$$

Define

$\Phi(t) = <\varphi, p_t>$ we get integrating by parts

$I_s = \int_0^{+\infty} \Phi(t)\; f_\Lambda^s\; (\frac{d}{dt}) k_\Lambda(t)\; dt;$ using then 1.3

$$f_\Lambda^s\; (\frac{d}{dt}) k_\Lambda(t) = \int_{-i\infty}^{+i\infty} \frac{f_\Lambda^s(\zeta)}{f_\Lambda(\zeta)}\; e^{\zeta t}\; \frac{d\zeta}{2i\pi}\ .$$

Now

$$||\Phi||_{L^\infty} \leq ||\varphi||_{L^\infty}$$

$\Phi^k(t) = <\Delta^k\varphi,\, p_t>$ and $||\varphi^k(t)||_{L^\infty} \leq ||\Delta^k\varphi||_{L^\infty}$

we shall integrate on $\operatorname{Re}\zeta = -\frac{1}{2}$

$$I_s = \int_0^{+\infty} \Phi(t)\; dt \int_{\operatorname{Re}\zeta=-\frac{1}{2}} \frac{f^s(\zeta)}{f_\Lambda(\zeta)}\; e^{\zeta t}\; \frac{d\zeta}{2i\pi}$$

$$= \int_{\operatorname{Re}\zeta=-\frac{1}{2}} \frac{f^s{}_\Lambda(\zeta)}{f{}_\Lambda(\zeta)}\; \frac{d\zeta}{2i\pi} \int_0^{+\infty} \Phi(t)\; e^{\zeta t}\; dt$$

$$\int_0^{+\infty} \Phi(t)\; e^{\zeta t}\; dt = \frac{\Phi(0)}{\zeta} - \frac{\Phi'(0)}{\zeta^2} + \frac{1}{\zeta^2} \int_0^{\infty} \Phi''(t) e^{\zeta t}\; dt$$

$$\frac{1}{2i\pi} \int_{\operatorname{Re}\zeta=-\frac{1}{2}} \frac{f_\Lambda^s(\zeta)}{f_\Lambda(\zeta)}\; \frac{d\zeta}{\zeta} = \frac{f_\Lambda^s(0)}{f_\Lambda(0)} \longrightarrow 1$$

$$\frac{1}{2i\pi}\int_{\mathrm{Re}\zeta=-\frac{1}{2}} \frac{f^s_\Lambda(\zeta)}{f_\Lambda(\zeta)}\ \frac{d\zeta}{\zeta^2}\ \left[\frac{d}{d\zeta}\left(\frac{f^s_\Lambda}{f_\Lambda}\right)\right]_{\zeta=0} \longrightarrow 0$$

As $|f^s_\Lambda(-\frac{1}{2}+i\tau)| < |f_\Lambda(-\frac{1}{2}+i\tau)|$ we can apply the Lebesgue dominated convergence for the last term and we get

$$\lim I_s = \Phi(0) + \int_{\mathrm{Re}\zeta=-\frac{1}{2}} \frac{d\zeta}{2i\pi\zeta^2}\int_0^\infty \Phi''(t)e^{\zeta t}\,dt$$

Now we can make a contour deformation and integrate on $\mathrm{Re}\zeta = -n$. Letting $n\to\infty$ we get that this last integral is zero and then the proposition is established.

6. Real analyticity outside the origin

6.1 Parabolic estimates

Let D a bounded domain in $R^m(\tilde{x}\in R^n)$; consider on $\tilde{D} = Dx[t_o,t_o+h]$ the parabolic operator

$$P = \frac{\partial}{\partial\tilde{t}} - \tilde{a}_{i,j}\ (\tilde{x},\tilde{t})\ \frac{\partial^2}{\partial\tilde{x}_i\,\partial\tilde{x}_j} - \tilde{b}_i\ (\tilde{x},\tilde{t})\ \frac{\partial}{\partial\tilde{x}^i}$$

where the coefficients have Hölder norm $\|\ \|_\alpha$, $\|\ \|_{1+\alpha}$ bounded:

$$\|a_{i,j}\|_\alpha < K_1, \quad \|b_i\|_{1+\alpha} < K_1$$

with the uniform ellipticity condition

$$\tilde{a}_{i,j}\ \xi^i\xi^j \geq K_2|\xi|^2 .$$

Suppose that the diameter of $\tilde{D}$ is bounded by K_1. Then there exists a constant M depending only on K_1, K_2, such that

$Pu = 0$ on $\tilde{D}$, $u \in C^2$ implies

$$\max_{\frac{1}{2}<t<1} \|u_t\|_{C^2_d(D)} \leq M\|u\|_{L^\infty(D\times|0,1|}$$

where $\|u\|_{C^2_d} = \max |\partial^i u| d + \max |\partial^{i,j} u| d^2$

d^2 denoting the distance of (x,t) to the boundary of $\tilde{D}$.

Proof: cf. [3] p. 92, Theorem 1.

6.2 Estimate of p_t in uniform norm

6.2.1 Lemma $\|p_t\|_{L^\infty(G)} < +\infty$ for $t > 0$.

Proof: We choose the domain D appearing in 6.1, as a domain of uni valence for the exponential chart

$$\xi \longrightarrow \exp(\xi)$$

$$D \longrightarrow G$$

Near a point $g_o \in G$ we will use the right exponential chart

$$\xi \longrightarrow \exp(\xi) g_o$$

Then the parabolic operator associated to $\frac{\partial}{\partial t} - \Delta$ in this chart will satisfy uniformly the requirement of 6.1.

We can apply the Harnack principle ([7] Theorem 2, p. 104)

$$C_1 < \frac{p_t(\exp(\xi)g)}{p_{\frac{t}{2}}(g)}$$

for $||\xi|| < \varepsilon$, where C_1, C_2 are independent of g. Let Θ be a function in $L^1(G)$ defined by

$$\Theta(\exp(\xi)) = \frac{1}{\mathrm{vol}(\{||\xi'||<\varepsilon\}} \quad \text{for } ||\xi|| < \varepsilon$$

and zero elsewhere. Then

$$||p_t * \Theta||_{L^\infty} \leq ||p_{t_1}||_{L^1}||\Theta||_{L^\infty} = ||\Theta||_{L^\infty}$$

By Harnack

$$||p_t * \Theta||_\infty > C_1||p_{\frac{t}{2}}||_{L^\infty}$$

which proves the lemma

6.2.2 <u>Lemma</u>: <u>Let</u> A <u>a compact set of</u> G, $e \notin A$. <u>Then there exist two constants</u> C_1, C_2 <u>such that</u>

$$||p_t||_{L^\infty(A)} < C_1 \exp\left(-\frac{C_2}{t}\right) \qquad (0 < t < 1)$$

<u>Proof</u>: The Varadhan [9] Malchanov [6] estimate gives:

$-t \log p_t(g)$ is equivalent to $\frac{1}{2}$ distance from e to g for the right invariant riemannian metric on G associated to Δ.

6.3 <u>Proposition</u>: $p_t(g)$ <u>is real analytic on any compact set, uniformly in</u> $t\varepsilon[a,+\infty[$, $a > 0$.

<u>Proof</u>:

$$||p_t||_{L^\infty} < ||p_{\frac{a}{2}}||_{L^\infty}||p_{t-a}||_{L^1} = ||p_{\frac{a}{2}}||_{L^\infty} = c, \ \forall t > \frac{a}{2}.$$

We shall take a basis $B_1 \ldots B_m$ of left invariant vector fields on G.

Given a multiindice $\alpha_1 \ldots \alpha_s$ $1 \leq \alpha_j \leq m$ we denote

$$B_\alpha = B_{\alpha_1} B_{\alpha_2} B_{\alpha_s} , \quad |\alpha| = s.$$

We want to prove that there exists a constant C

$$\| B_\alpha p_t \|_{L^\infty(G)} \leq s\,!\, c^s \qquad \forall \alpha \text{ with } |\alpha| = s.$$

We have by 6.1.

$$\| B_\alpha p_t \|_{L^\infty} \leq C \qquad \forall\, t > a, \ |\alpha| \leq 2.$$

We shall evaluate $B_\alpha p_t$ for $|\alpha| = 2n$.

We denote

$$B_\alpha p_t = p_t^\alpha , \quad \text{then}$$

$$\frac{\partial}{\partial t} p_t^\alpha - \Delta p_t^\alpha = 0$$

p_t^α satisfies the same estimates as p_t. We shall take a sequence of domain $\tilde{D}_1 \supset \tilde{D}_2 \supset \tilde{D}_k \supset \tilde{D}_n$ where we denote $t_k = t - \frac{1}{2} + \frac{k}{4n}$, $D_k = |t_k, t| xD$.

Then by 6.1.

$$\sup_{\substack{t_{k+1} < t < t \\ |\alpha| = k}} \| p^\alpha \|_{L^\infty(G)} \leq C n^2 \sup_{\substack{t_k < t < t \\ |\beta| = k-1}} \| p^\beta \|_{L^\infty(G)} .$$

Finally we get

$$\| p_t^\gamma \|_{L^\infty(G)} \leq n^{2n} C^n \qquad \text{for } |\gamma| = 2n .$$

6.4 Proposition: Let A a compact of G, $e \notin A$. then p_t belongs to the Gevrey class 2 on A, uniformly in $t \in]0,+\infty[$.

Proof: We use the same estimation as in 6.4 splitting the interval 0,t in the sub intervals

$$\frac{t}{2}(1+\frac{1}{n}),\ \frac{t}{2}(1+\frac{2}{n}),\ldots,\ \frac{t}{2}(1+\frac{n-1}{n}),\ t.$$

Then we get, denoting by A' a compact $A' \supseteq A$, $e \notin A'$ and using a sequence of approximating domains $A' \supset A_k \supseteq A_{k+1}$, $A_n = A$ such that

$$\text{distance } (\partial A_k, \partial A_{k+1}) = \frac{c}{n}$$

we get

$$||p^{\alpha}\cdot d^2_{D_{k-1}}||_{L^{\infty}(D_{k-1})} \leq ||p^{\beta}||_{L^{\infty}(D_{k-1})} \qquad \text{with: } |\alpha| = 2k, \quad |\beta| = 2(k-1)$$

$$D_k = [\tfrac{t}{2}(1+\tfrac{k}{n}),\ t] \times A_k.$$

By the approximation properties on the boundary we have

$$||p^{\alpha}\cdot||_{L^{\infty}(D_k)} \leq (\frac{2\pi}{t})^2 ||p^{\alpha}\cdot d^2_{D_{k-1}}||_{L^{\infty}(D_{k-1})}.$$

Finally

$$||p^{\alpha}_t||_{L^{\infty}(A)} \leq (\frac{2\pi}{t})^{2n} \exp(-\frac{c}{t}) \qquad \text{for } |\alpha| = 2n.$$

As $\sup\limits_{o<t<1} t^{-2n} \exp(-\frac{c}{t}) = n^{2n} c^n$ we get

$$\sup_{o<t<1} ||p^{\alpha}_t|| \leq n^{4n} c^n \quad \text{with } |\alpha| = 2n.$$

7. Final proofs.

7.1 Proof of the Theorem

2.2. is a consequence of 4.1. and 3.4.

2.3. is a consequence of 6.3., 6.4. and 4.2.

2.1. is a consequence of 4.2.

7.2 Proof of the factorization at step two (2.4.)

We follow exactly the method of [1] ; given $\varphi \varepsilon$ (G) we can find Λ such that for all $R < +\infty$

$$\sum_{n=0}^{+\infty} R^n a_{n,\Lambda} ||\varphi||_{C^n(G)} < +\infty .$$

Let $h \varepsilon \mathcal{D}(G)$ belonging to the Gevrey class of order 2, J_2 and such that $h \equiv 1$ on a neighborhood of e.

Define

$$q_1 = q_\Lambda h .$$

Then

$$L^s_\Lambda(q_1) = h L^s_\Lambda (q_\Lambda) + w_s$$

where w_s converges in J_2 when $s \to +\infty$ to $w_\infty \varepsilon J_2$

$$L^s_\Lambda (q_1 * \varphi) (x) = \int (L^s_\Lambda q_\Lambda) (y) h(y)\varphi(y^{-1}x) dy + w_s * \varphi$$

Denote $r(y) = h(y)\varphi(y^{-1}x)$ and apply 5.1. then

$$\int (L^s q_\Lambda)(y) r(y) dy \to r(e) = \varphi(x) .$$

Therefore

$$\lim_{s\to\infty} L_\Lambda^s (q_1 * \varphi) = \varphi + w_\infty * \varphi \quad \text{where} \quad w_\infty \varepsilon \mathcal{J}_2$$

Denote $\tilde{\Phi}(x) = \varphi(x^{-1})$

$$\begin{aligned} L_\Lambda^s (q_1 * \varphi)(x) &= \int q_1(y)\; L_\Lambda^s (\varphi(y^{-1}x))dy \\ &= \int q_1(y)\; L_\Lambda^s (\varphi(x^{-1}y))dy \\ &= \int q_1(y)(L_\Lambda^s \tilde{\Phi})(x^{-1}\; ydy) \\ &= \int q_1(y)\; \widetilde{L_\Lambda^s \tilde{\Phi}}(y^{-1}x(dy) \\ &= q_1 * \widetilde{L_\Lambda^s \tilde{\Phi}} \end{aligned}$$

As $x \to x^{-1}$ is an analytic diffeomorphism it conserves the class $\mathcal{D}_{M_n}$ therefore when $s \to \infty$, $L_\Lambda^s \tilde{\Phi}$ converges in $\mathcal{D}$ to a function w' and we have

$$\varphi = q_1 * q' - w_\infty * \varphi, \text{ with } q' = \tilde{w}'$$

BIBLIOGRAPHY

[1]. J. Dixmier et P. Malliavin: Factorisations de fonctions et de vecteurs indéfiniment différentiables. Bulletin des Sciences Math. 102 (1978) p. 305-330.

[2]. J. Faraut: Calcul symbolique sur les générateurs infinitésimaux de semi groupes d'opérateurs. Annales de Fourier 20 (1978) p. 235-301.

[3]. A. Friedmann: Parabolic Equation. Academic Press, 1967.

[4]. L. Gärding: Vecteurs analytiques dans les représentations des groupes de Lie. Bulletin Société Mathématiques de France 88 (1960) p. 73-93.

[5]. A. Hulanicki: Subalgebra of $L_1(G)$ associated with Laplacian on a Lie group. Coloqu. Mathematic, vol. XXXI 1974, p. 259-285.

[6]. Molchanov: Diffusion processes and Riemannian Geometry. Uspehi Mat. Nauk 1975 p. 1-59.

[7]. J. Moser: A Harnack inequality for parabolic equation. Comm. Pure and Applied Mathematics 17 (1964) p. 101-134.

[8]. L.A. Rubel, W.A. Squires and B.A. Taylor: Irreducibility of certain entire functions with applications to Harmonic Analysis. Annals of Mathematics 1978 (108) p.553-567.

[9]. S.R.S. Varadhan: Diffusion process on small time interval. Comm.Pure and Applied Math. 20 (1967) p. 659-685.

DISTANCE AND VOLUME DECREASING THEOREMS FOR A FAMILY OF HARMONIC MAPPINGS OF RIEMANNIAN MANIFOLDS

N.C. Petridis

1. The main results

The classical lemma of Schwarz states:

"Let $f: D \to D$ be a holomorphic mapping of the unit disc to the unit disc such that $f(0) = 0$. Then $|f(z)| \leq |z|$ and $|f'(0)| \leq 1$".

Introducing the Poincaré-Bergman metric for D

$$dS^2 = \frac{d\,z\,d\,\bar{z}}{(1-|z|^2)^2}$$

Schwarz's lemma becomes equivalent to:

"If $f: D \to D$ is holomorphic then

$$f^*dS^2 \leq dS^2 \text{ ",}$$

where f^* is the pull back of differential forms. We notice that the Poincaré-Bergman metric makes D a Kaehler manifold of constant curvature-4. With this remark, we are led in a natural way, to the following extension of Schwarz's lemma, given by Ahlfors: "Let M be a one-dimensional Kaehler manifold with metric dS_M^2 whose Gaussian curvature is bounded above by a negative constant-B. Let D_a be an open disc of radius a with metric given by

$$dS_D^2 = \frac{4\,a^2\,dz\,d\bar{z}}{A(a^2-|z|^2)^2}$$

(which makes D_a a Kaehler manifold of constant curvature $-A < 0$). If $f: D_a \to M$ is holomorphic then

$$f^*\,dS_M^2 \leq \frac{A}{B}\,dS_D^2 \text{ " .}$$

If $A < B$ the mapping is distance decreasing. In the general case we will say that the mapping is distance (or volume) decreasing up to a constant.

Chern extended the lemma to holomorphic mappings between complex manifolds of higher dimensions. The lemma was further extended, in the form of distance and volume decreasing theorems, by S. Kobayashi [7] and, more recently, by S.T. Yau [10]. Real analogues and generalizations of the lemma (as well as Liouville's and Picard's first theorem) have recently been obtained for harmonic mappings [2], quasiconformal mappings [5], and for mappings of bounded dilatation [4].

In all those extensions of the Schwarz - Ahlfor's lemma there is a variation of generality depending on the conditions imposed on the domain space. The most general domain considered so far is that of Yau's [10], which is a complete Riemannian manifold with Ricci curvature bounded from below. The target space has been, almost invariably, assumed to be hyperbolic. In a recent paper [9] the author extended Picard's first theorems to mappings with target space not necessarily hyperbolic. In this note, taking advantage of a Yau's condition, the Schwarz - Ahlfors lemma is extended, in the form of distance and volume decreasing theorems, to mappings with target space not necessarily hyperbolic. The main results:

<u>Theorem: Let M be an m-dimensional complete Riemannian manifold with Ricci curvature bounded from below, and let N be an n-dimensional Riemannian manifold with sectional curvatures bounded from above by some constant H and scalar curvature bounded from above by a negative number $-C < 0$, related to H by $H < C/m(m-1)(K^4-1)$. If $f: M \to N$ is an harmonic K-quasiconformal mapping then, if $n \leq m$ the mapping is distance decreasing and if $n = m$ the mapping is volume decreasing (up to a constant).</u>

<u>Corollary 1. Let</u> M <u>be a 2m-dimensional complete quasi-Kaehlerian manifold with Ricci curvature bounded below and let</u> N <u>be a 2n-dimensional quasi-Kaehlerian manifold with curvature conditions as in the theorem.</u>

<u>If</u> $f: M \to N$ <u>is an almost complex K-quasiconformal mapping then, for</u> $m = n$ <u>the mapping is volume decreasing, for</u> $n \leq m$ <u>the mapping is distance decreasing.</u>

If in (13) we have $B = 0$, we get $||f_*||^2 \equiv 0$, which proves the following Picard's type theorem:

<u>Corollary 2. Let</u> M <u>be an m-dimensional complete Riemannian manifold with non-negative Ricci curvature and let</u> N <u>be as in the theorem. Then if</u> $f: M \to N$ <u>is a K-quasiconformal harmonic mapping it is reduced to a constant.</u>

In the following two examples we can see that the restriction $n \leq m$ on the dimension of the manifolds and the condition that the scalar curvature must be bounded away from zero cannot be dropped.

<u>Example 1</u>. Let $\mathbb{C}$ be the complex plane with flat metric $ds_1^2 = |dz_1|^2$, where z_1 is the complex coordinate in $\mathbb{C}$, and let $\varphi(z)$ be an arbitrary holomorphic function. The mapping $\varphi(z)$ is not, in general, expected to be distance decreasing. Let $\mathbb{C}^* = C-\{0,1\}$ with coordinate z_2 and Kaehler metric ds_2 with Gauss curvature bounded above by -4 [7]. We consider the manifold $N = \mathbb{C} \times C^*$ with the product metric $ds^2 = ds_1 x ds_2$. Then the scalar curvature of N is bounded above by -4. We consider the mapping $f = \mathbb{C} \to N$, defined by $z_1 = \varphi(z)$, $z_2 = \text{const}$. This mapping is not, in general, distance decreasing.

<u>Example 2</u>. Let $M = \mathbb{C}$ with the flat metric and $N = \mathbb{C}$ with the conformal metric $ds^2 = (1+|z|^2)dzd\bar{z}$. The curvature of N is strictly

negative everywhere, but not bounded away from zero. The identity mapping is trivially conformal but not distance decreasing.

2. Harmonic mappings and curvature

Let M and N be C^∞ oriented Riemannian manifolds of dimensions m and n respectively, and let $f : M \to N$ be a smooth mapping. If ds_M^2 and ds_N^2 are the metrics of M and N respectively, the inverse image $f^*ds_N^2$ is a positive semidefinite form in M and its trace $Tr(f^*ds_N^2)$, relative to ds_M^2, is a function in M taking nonnegative values. If D is a compact domain in M, the energy of f over D is defined by the integral

$$E(f,D) = \frac{1}{2} \int_D Tr(f^*ds_N^2)\, dV_M$$

where dV_M denotes the volume element of M. Thus $E(f,D)$ can be considered as the generalization of the classical Dirichlet integral.

The mapping f is defined to be harmonic over D if $E(f,D)$ has a critical value relative to all mappings which agree with f on the boundary ∂D of D. We shall say that f is a harmonic mapping if it is harmonic over any compact domain of M.

In this section, instead of this "variational" point of view of harmonicity, we shall make use of a necessary and sufficient condition for a mapping to be harmonic, derived by J. Eells and J.H. Sampson [3]. To set the terminology, we recall some of the necessary Riemannian geometry involved in the discussion.

Using the method of moving frames, we have locally

$$ds_M^2 = \sum_{i=1}^{m} \omega_i^2 \quad \text{and} \quad ds_N^2 = \sum_{a=1}^{n} \omega_a^{*2}$$

where ω_i and ω_a^* are linear differential forms in M and N

respectively. The structure equations in M are

$$d\omega_i = \sum_j \omega_j \wedge \omega_{ji}, \quad \omega_{ij} + \omega_{ji} = 0$$

$$d\omega_{ij} = \sum_k \omega_{ik} \wedge \omega_{kj} + \Omega_{ij}, \quad \Omega_{ij} + \Omega_{ji} = 0,$$

where ω_{ij} are the connection forms and Ω_{ij} are the curvature forms given, in terms of the components $R_{ij,kl}$ of the curvature tensor, by

$$\Omega_{ij} = -\frac{1}{2} \sum_{k,l} R_{ijkl}\, \omega_k \wedge \omega_l$$

The Ricci tensor R_{ij} is defined by

$$R_{ij} = \sum_k R_{ikjk}$$

and the scalar curvature R by

$$R = \sum_i R_{ii}$$

Corresponding quantities in N will be denoted with an asterisk. Under the mapping f a tensor field with components A_i^a is defined in M, by

$$f^* \omega_a^* = \sum_i A_i^a \omega_i$$

where f^* is the pull-back map (which in the sequel will be dropped when its presence is clear from the context).

The covariant differential of the tangent map f_* is defined by

(1)
$$DA_i^a \equiv dA_i^a + \sum_j A_j^a \omega_{ji} + \sum_b A_i^b \omega_{ba}^* \equiv \sum A_{ij}^a \omega_j \text{ (say)}$$
$$A_{ij}^a = A_{ji}^a$$

The tensor field with components

$$\tau_\alpha = \sum_i A^a_{ii}$$

is the tension vector field of J. Eells and J.H. Sampson who proved [3] that the mapping f is harmonic if and only if

$$\tau_\alpha = 0.$$

Examples.

1. If $f = (f_1, f_2, \ldots, f_n)\colon E^m \to E^n$ (Euclidean spaces) is a smooth mapping the Eells-Sampson tensor becomes:

$$\tau_\alpha = \sum \frac{\partial^2 fa}{\partial x_i^2}$$

2. If M and N are Kähler manifolds, then every holomorphic $M \to N$ is harmonic, with respect to any compatible metric.

Now, we wish to calculate the Laplacian of the ratio of the line elements $f{*}dS_N/dS_M = ||f_*||$.

Differentiating (1) and using the structure equations we get

$$\text{(2)} \qquad \sum DA^a_{ij} \wedge \omega_j = \sum A^a_j \,\Omega_{ji} + \sum A^b_i \,\Omega^*_{ba}$$

where

$$DA^a_{ij} = dA^a_{ij} + \sum_k A^a_{kj}\,\omega_{ki} + \sum_k A^a_{ij}\,\omega_{kj} + \sum_b A^b_{ij}\,\omega^*_{ba}$$

$$= \sum_k A^a_{ijk}\,\omega_k \quad \text{(say)}.$$

For a C^∞ function φ on M the Laplacian $\Delta\varphi$ is defined in terms of the covariant differential ∇ in M by

$$\Delta\varphi = \sum_k \nabla^2\varphi(e_k, e_k)$$

where $\{e_i\}$ is the orthonormal frame dual to $\{\omega_i\}$. Applying this formula to

$$\varphi = ||f_*||^2 = \sum_{a,i} (A_i^a)^2$$

and using (1), (2), (3), as in [4], we get

$$\frac{1}{2}\Delta||f_*||^2 = \sum_{a,i,j} (A_{ij}^a)^2 + \sum_{a,i,j} R_{ij} A_i^a A_j^a$$

$$- \sum_{\substack{a,b,c,d\\ i,j}} R^*_{abcd} A_i^a A_j^b A_i^c A_j^d + \sum_{a,i,j} A_i^a A_{jji}^a$$

If f is harmonic, the last terms in (4) vanishes and we obtain the real analogue of the Chern-Lu formula:

$$(5) \quad \frac{1}{2}\Delta||f_*|| \geq \sum_{a,i,j} R_{ij} A_i^a A_j^a - \sum_{\substack{a,b,c,d\\ i,j}} R^*_{abcd} A_i^a A_j^b A_i^c A_j^d$$

3. Mappings of Bounded Dilation

The notation being as is the previous section, let $f_*: T_x(M) \to T_{f(x)}(N)$ be the tangent mapping and h a metric in N. The tensor f^*h is a symmetric semidefinite quadratic form on $T_x(M)$. With a suitable choice of frames in $T_x(M)$ and $T_{f(x)}(N)$, this quadratic form becomes

$$f^* h = \sum_{i=1}^k \gamma_i^2 \, \omega_i \otimes \omega_i \quad \text{at} \quad x,$$

where k is the rank of f_* and the positive numbers γ_i^2, $i = 1,2,\ldots,k$ are the eigenvalues of f^*h. The ratio $l_i = \gamma_i/\gamma_{i+1}$ is called the ith dilatation of $f_*(x)$. If the first dilatation is bounded over M the mapping is called of bounded dilatation [4]. The mapping is called K-quasiconformal if the last dilatation l_{k-1} is bounded over M by a constant K [5]. Of course every quasiconformal mapping is of bounded dilatation. An almost complex mapping is also of bounded dilatation.

These definitions can be given a natural geometrical interpretation. The tangent mapping f_* maps a (k-1)-sphere to a (k-1) ellipsoid. The lengths of the principal axes of this ellipsoid are $\gamma_1 \geq \gamma_2 \ldots \geq \gamma_k > 0$ and $l_1 = \frac{\gamma_1}{\gamma_2}$, $l_{k-1} = \frac{\gamma_1}{\gamma_k}$.

We have the following proposition [4].

<u>Proposition</u> 1 <u>A</u> C^∞ <u>mapping</u> $f : M \to N$ <u>is of bounded dilatation if and only if</u>

$$||f_*||^2 \leq k\,K^2 ||\Lambda^2 f_*||$$

<u>where</u> $k = \min(m,n)$, K <u>is an upper bound of the dilatation</u> l_1 <u>and</u> $||\Lambda^2 f_*||^2 = \sum_{i_1<i_2} \gamma_{i_1}^2 \gamma_{i_2}^2$

If $f_*(e_i) = A_i$, as in the previous section, the mapping $f_*\colon \Lambda^p T_x(M) \to \Lambda^p T_{f(x)}(N)$ is defined by

$$f_*(e_{i_1} \Lambda \ldots \Lambda e_{i_p}) = f_*(e_{i_1}) \Lambda \ldots \Lambda f_*(e_{i_p})$$

and with the induced metrics in Λ^p we have

$$||\Lambda^p f_*||^2 = \text{trace } \Lambda^p ({}^t f_* f_*) = \sum_{1<i_1 \ldots <i_p<m} \gamma_{i_1}^2 \ldots \gamma_{i_p}^2$$

where ${}^t f_*$ is the transpose of f_*.

In particular,

$$||\Lambda^2 f_*||^2 = \sum_{i<j} ||A_i \Lambda A_j||^2$$

For a K-quasiconformal mapping we have the following inequalities [4]

$$\left[\frac{||\Lambda^q f_*||^2}{\binom{k}{q}}\right]^{1/q} \leq \left[\frac{||\Lambda^p f_*||^2}{\binom{k}{p}}\right]^{1/p} \leq K^2 \left[\frac{||\Lambda^q f_*||^2}{\binom{k}{p}}\right]^{1/q}$$

The following inequality, which follows directly from the definition of a K-quasiconformal mapping, will be used in the next section

(7) $$||f_*||^2 < nK^2 \, ||A_i|| \quad ||A_j||$$

for any i,j.

4. Distance Decreasing Theorems

In [4] the following theorem was established:

<u>Proposition 2</u>. <u>Let</u> M <u>be a complete connected Riemannian manifold of constant negative curvature</u> -A <u>and let</u> N <u>be a Riemannian manifold whose sectional curvatures are bounded above by a negative constant</u> -B. <u>If</u> $f: M \to N$ <u>is a harmonic mapping of bounded dilatation of order</u> K, <u>then the following inequality is satisfied</u>:

$$||\Lambda^p f_*||^{2/p} < k \binom{k}{p}^{1/p} \frac{m-1}{2} \frac{A}{B} K^4 \qquad 1 < p < k.$$

For $p = 1$ we have a distance decreasing property and for $p = m = n = k$ we have a volume decreasing property.

Modifying the proof of this proposition and using a proposition of S.T. Yau (see below), Proposition 2 can be substantially improved [6] by assuming M to be a <u>complete connected Riemannian manifold with Ricci curvature bounded from below</u>.

Now, we proceed to prove the theorem announced in 1.

For the proof of this theorem we need the following Proposition due to S.T. Yau [10].

<u>Proposition 3</u>. <u>Let</u> u <u>be a</u> C^2 <u>function defined and bounded from below. Then there exists a sequence of points</u> $\{q_n\} \subset M$ <u>such that</u>.

$$\lim_{n\to\infty} u(q_n) = \operatorname{Sup} u , \quad \lim_{n\to\infty} ||du(q_n)|| = 0, \quad \lim_{n\to\infty} \operatorname{Sup} \Delta u(q_n) \leq 0$$

With the notation as in the last section, let

$$u = ||f_*||^2 = \sum_{a,i} (A_i^a)^2$$

The function $F = -\frac{1}{\sqrt{u+1}}$ is bounded on M. We observe that when F tends towards its Supremum then u also tends towards its sup. Straight computation shows that the Laplacians of u and F are related by

$$\frac{1}{6}(F)^{-2}\Delta u = \frac{1}{3}(F)^{-1/2}\,\Delta F + ||df||^2$$

If we assume that M is as in Proposition 3 then this proposition is applicable on F and we have

<u>Lemma 1</u>. <u>Let</u> M <u>be a complete Riemannian manifold with Ricci curvature bounded from below and let</u> u <u>be a non negative</u> C^2 <u>function on</u> M. <u>There is a sequence of points</u> $\{q_n\} \subset M$ <u>such that</u>

$$\lim_{n\to\infty} \operatorname{Sup}[u(q_n)+1]^{-2} \Delta u(q_n) \leq 0$$

$$\lim_{n\to\infty} u(q_n) = \operatorname{Sup} u$$

Proof of the Theorem. Let $f:M\to N$ be as in the theorem. At a point x where $f_* \neq 0$, a frame $\{e_1, e_2, \ldots, e_m\}$ of M can be chosen so that $A_i \perp A_j$, for $i \neq j$, where $A_i = f_* e_i$ and $i = 1, 2, \ldots, n$. With such an arrangement, the sectional curvature of the section spanned by A_i, A_j is given by

$$H_{ij} = \frac{\sum R^*_{abcd} A_i^a A_j^b A_i^c A_j^d}{||A_i||^2 \, ||A_j||^2}$$

so, we have

$$(8) \qquad \sum R^*_{abcd} A_i{}^a A_j{}^b A_i{}^c A_i{}^c A_j{}^d = \sum H_{ij} ||A_i||^2 ||A_j||^2$$

If H'_{ij} and H''_{ij} are defined by

$$H'_{ij} = \frac{H_{ij} + |H_{ij}|}{2}, \quad H''_{ij} = \frac{H_{ij} - |H_{ij}|}{2}$$

and $||A_1||$, $||A_n||$ are the max and min respectively of $||A_i||$ for $i = 1, 2, \ldots, n$ (this can be assumed without loss of generality), we obtain from (8)

$$(9) \qquad \frac{1}{2} \sum R^*_{abcd} A_i^a A_j^b A_i^c A_j^d \leq (\sum H'_{ij}) ||A_1||^4 - (\sum |H''_{ij}|) ||A_n||^4$$

$$= \left[(\sum H'_{ij}) \frac{||A_1||^4}{||A_n||^4} - \sum |H''_{ij}| \right] ||A_n||^4$$

Assuming that the mapping is K-quasiconformal (i.e. $\frac{||A_i||}{||A_n||} < K$) that $H'_{ij} < H$ and that $\sum H'_{ij} - \sum |H''_{ij}| < -C$, we obtain from (9)

$$(10)\qquad \sum R^{*}_{abcd}\, A^a_i\, A^b_j\, A^c_i\, A^d_j \leq 2\{(m(m-1)(K^4-1)\ H-C\}||A_n||^4$$

If we put $A = C-m(m-1)(K^4-1)H$, then from the assumption on H, (10) becomes

$$\sum R^{*}_{abcd}\, A^a_i\, A^b_j\, A^c_i\, A^d_j \leq -2A||A_n||^4$$

Finally, from (7) we obtain

$$(11)\qquad \sum R^{*}_{abcd}\, A^a_i\, A^b_j\, A^c_i\, A^d_j \leq -2A\ \frac{||f_*||^4}{n^2K^4}$$

Now, we take into account the assumption that the Ricci curvatures of M are bounded below by a constant-B. Hence

$$\sum R_{ij}\, A^a_i\, A^a_j \geq -B||f_*||^2$$

Reminding, finally, that f is harmonic, we have from (5), (11) and (12)

$$\frac{1}{2}\,\Delta||f^2_*||^2 \geq -B||f_*||^2 + A\ \frac{||f_*||^4}{n^2K^4}$$

Applying Lemma 1, there is a sequence of points $\{q_n\} \subset M$ such that

$$\limsup_{n\to\infty} \frac{||f_*||^2(q_n)\,[-B + \frac{A}{n^2K^4}\ ||f_*||^2(q_n)]}{[\,||f_*||^2(q_n) + 1]^2} \leq 0$$

$$\lim_{n\to\infty}||f_*||^2(q_n) = \sup||f_*||^2$$

from which we obtain

$$\text{(13)} \qquad \sup||f_*||^2 \leq \frac{B}{A} n^2 K^4$$

This inequality proves the theorem for the distance decreasing property. As far as the volume decreasing property,is concerned comparing (6) and (13) we obtain.

$$||\Lambda^n f_*||^2 \leq \sqrt{n} \left|\frac{B}{A} n^2 K^4\right|^{1/n}$$

which completes the proof of the theorem.

5. The Hermitian case

Let M be a 2m-dimensional almost complex manifold with complex structure J and Riemannian metric g. If the metric g is invariant by J, i.e. $g(JX,JY) = g(X,Y)$ for any vector fields X and Y the metric is called Hermitian and the manifold is called almost Hermitian. A linear connection on M such that the metric tensor and the complex structure are parallel is called Hermitian connection with pure torsion tensor T, i.e. $T(JX,X) = T(X,JY)$ for all vector fields X and Y on M, is called canonical connection. The existence and uniqueness of the canonical connection is assured by the general theory [1]. Let Φ be the Kaehler form of M, i.e. $\Phi(X,Y) = g(X,JY)$ for all vector fields X and Y. If the Kaehler form Φ is closed M is called almost Kaehlerian manifold and if the part of $d\Phi$ of bidegree (1,2) vanishes M is called quasi Kaehlerian.

It is well known that a holomorphic mapping of Kaehlerian manifolds is harmonic. This has been extended by Lichnerowitz [8]: Proposition for quasi Kaehlerian manifolds M and N every almost complex mapping $f: M \to N$ is harmonic in terms of the corresponding metrics. Corollary 1 is an immediate consequence of this proposition and the Theorem.

REFERENCES

[1]. S.S. Chern, Characteristic classes of hermitian manifolds. Ann. of Math. 47 (1946) 85-121.

[2]. S.S. Chern and S.I. Goldberg, On volume decreasing property of a class of real harmonic mappings. Am.J.Math. 97 (1975) 133-147.

[3]. J. Eells, Jr. and G.H. Sampson, Harmonic mappings of Riemmanian manifolds.

[4]. S.I. Goldberg, T. Ishihara and N.C. Petridis, Mappings of bounded dilatation of Riemannian manifolds. J.Diff.Geom. 10 (1975) 619-630.

[5]. S.I. goldberg and T. Ishihara, Harmonic quasiconformal mappings of Riemmanian manifolds. Bull. Am.Math.Soc. 80 (1974) 562-566.

[6]. S.I. Goldberg and Zvi Har'El, A general Schwarz lemma for Riemannian manifolds. To appear.

[7]. S. Kobayashi, Hyperbole manifolds and holomorphic mappings. M. Dekker (1970).

[8]. A. Lichnerowicz, Applications harmoniques et variétés Kahlériennes. Symposia Mathematica 3, Bologna (1970) 341-402.

[9]. N.C. Petridis, A generalization of the little theorem of Picard. Proc. Amer. Math. Soc. 61 (1976) 265-271.

[10]. S.T. Yau, A general Schwarz lemma for Kahler manifolds. Am.J. of Math. v. 100,

East Illin. University
Charleston Illinois U.S.A.
University of Crete
Iraklion Crete,Greece

ON THE L' NORM OF EXPONENTIAL SUMS

S.K. Pichorides

N.R.C. Demokritos, Athens

In this talk we shall give a review of the known results concerning the conjecture of Littlewood on exponential sums and a brief sketch of a proof of the best known results in this direction up to date.

1. <u>Origin of the problem</u>. It is now about 50 years that Hardy and Littlewood in their effort to characterize those functions of L^q, $q > 1$, which remain in L^q after changing the order and argument of their Fourier coefficients, they obtained the following remarkable inequality:

(1) $||1 + 2\cos n_1 x + \ldots + 2\cos n_N x||_q \leq ||1 + 2\cos x + \ldots + 2\cos Nx||_q$,

$(n_1, n_2, \ldots, n_N$ are distinct positive integers) $\qquad q = 2,4,\ldots$

In fact the argument they gave was not conclusive. A few years later the above inequality, which is a special case of a more general one known as the Hardy-Littlewood rearrangement theorem, was slightly extended and correctly proved by Gabriel.

It appears that it is still unknown if (1) holds for all $q \geq 2$ (i.e. not necessarily for even integers). What is of interest to us now is the possible extensions of (1) for $q < 2$ and above all for $q = 1$.

Let's write $f(x) = 1 + 2\cos n_1 x + \ldots + 2\cos n_N$ and $g(n) = 1 + 2\cos x + \ldots + 2\cos Nx$. Then (1) becomes $||f||_q \leq ||g||_q$, $q = 2,4,\ldots$. If we plot the graphs of $||f||_q^q$, $||g||_q^q$, as functions of q we obtain two logarithmically convex curves which intersect at $q = 2$ and the curve $||g||_q^q$ stays below $||g||_q^q$ when $q = 4,6,\ldots$

This makes the conjectures $||f||_q \leq ||g||_q$, $q \geq 2$ and $||f||_q \geq ||g||_q$, $q \leq 2$, in particular $||f||_1 \geq ||g||_1$, at least plausible.

A slightly weaker version of $||f||_1 \geq ||g||_1$ is known as

Littlewood's conjecture. It appeared explicitly for the first time in a joint paper by Hardy and Littlewood in 1948. In this paper Hardy and Littlewood commenting on (1) said that it was obtained after 4 years of fruitless efforts (!).

The slightly weaker version we mentioned above reads as follows:

(2) $$||\exp(in_1x)+\ldots+\exp(in_Nx)||_1 \geq C \log N$$

where C is an absolute positive constant. Even this weaker form despite 30 years of efforts has not been decided yet.

2. <u>The problem of the one sided L^∞ norm</u>. A problem closely related to (2) was posed by Askey and Chowla in connection with some number theoretic questions.

We write:

$$F(x) = \exp(in_1x) +\ldots+ \exp(in_Nx) = (\cos n_1x+\ldots+\cos n_Nx) +$$

$$+ i(\sin n_1x +\ldots+ \sin n_Nx) = f(x) + i\tilde{f}(x)$$

and $M = |\min f(x)|$.

We observe that $M + f \geq 0$ and hence

$$M = \int (M+f) = \int |M+f| \geq ||f||_1 - M$$

This implies

$$M \geq \frac{1}{2} ||f||_1$$

Thus, any lower bound of $||f||_1$ gives automatically a lower bound for M. Here the conjectured lower bound is of the order of $\sqrt{N}$ and not logN. However, the best bound we know for M is the same as the one we know for $||f||_1$ (which we obtain trivially by the above argument).

3. Historical review. We shall mention only the results concerning the general case (i.e. without any essential restrictions on the frequencies $n_1,\dots,n_N$) although some results concerning special sequences $n_1,\dots,n_M$ were obtained by very interesting arguments (e.g. by Salem, Selberg (for the one sided L^∞ norm), and Dixon . The lower bounds for $||F||$ known up to 1977 were of the form (logN/loglogN) .

The first significant result ($\alpha = 1/8$) towards the solution of Littlewood's conjecture was obtained by P. Cohen (1959). Davenport (1960) and Pichorides (1974) improved the exponent $\alpha\alpha$ to 1/4 and 1/2 respectively. Roth in 1973 obtained the same bound ($\alpha = 1/2$) for M. The above results were based on more or less the same underlying idea (with the possible exception of Roth's result).

Now, we know that $||F|| = C(\log N)^{1/2}$. This was proved in 1977 by Pichorides. The proof was based on an apparently completely different kind of argument which we shall sketch in the next paragraph. A few months ago another proof of the same result was obtained by Fournier. Fournier proved a more general result which reads as follows:

"If E is an infinite bounded below set of integers then there is a strictly increasing sequence $\{h_k\}_{k=0}$ of elements of E such that for each index k there are fewer than 4^k elements of E less than h_k and such that

$$\sum_{k=0}^{\infty} |f(h_k)|^2 \le 8(||f||_1)^2$$

for all functions f with $\hat{f}(m) = 0$ if $m \notin E$.

This result (from which the inequality $||F||_1 \ge C\sqrt{\log N}$ follows trivially) has to be compared with the classical Paley theorem on H^1 functions.

4. Outline of the proof of $||F||_1 \ge C(\log N)^{1/2}$

Roughly speaking the proof proceeds as follows: We assume, as we may, that there are more even, say N_e, than odd, say N_o, frequencies n_i. We prove that the norm $||F||_1$ that we want to estimate, exceeds the average of the norms corresponding to the exponential sums of the odd and even frequencies by at least $C(\log N)^{-1/2}$. This implies the desired result if $N_o \geq \frac{1}{2} N_e$.

If $N_o < (N_e/2)$ then $||F||_1$ exceeds the norm corresponding to the odd frequencies by a fixed positive quantity. This again concludes the proof except in the case of very small N_o $(<N/(\log N)^2)$.

In the last case $(N_o < N/(\log N)^2)$ we repeat the argument with the exponential sum corresponding to the even frequencies. Continuing this way either we get the desired result or we obtain a long enough sequence of frequencies consisting of odd multiples of distinct powers of 2. This case can be settled with the help of standard methods of the theory of Fourier series.

In order to carry out this program we translate the sequence of frequencies $n_1, n_2, \ldots, n_N$ in such a way that:

(i) There are positive integers $k_1 < k_2 < \ldots < k_t$ such that every n_k, $k=1,2,\ldots,N$ is an odd multiple of 2^{k_r} for some r, $1 \leq r \leq t$ and

(ii) The number of even multiples of 2^{K_r} is not less than the number of the odd ones.

Now, using induction we try to obtain our estimate from the known estimates about the L' norms of the exponential sums corresponding to the even and odd frequencies or if this is not possible to obtain the long sequence of distinct multiples of 2 mentioned above.

In the last case we conclude the proof by using the following:

<u>Lemma</u> - "Suppose that $m_1, m_2, \ldots$ is a sequence of positive integers such that any integer can be written in at most one way in the form $b_1m_1 + \ldots + b_nm_n$ for some n, where $b_i \varepsilon \{-1,0,1\}$, $i=1,2,\ldots$. Then for any $g(x) = a_o + a_1\exp(ix) + \ldots + a_n\exp(inx) + \ldots$ we have

$$\left(\sum_{n=1}^{\infty} |\alpha_{m_n}|^2\right)^{1/2} \leq C \int |g| (\log^+ |g|)^{1/2} + C$$

where $\log^+\alpha = \log\alpha$ is $\alpha > 1$ and 0 if $0 < \alpha \leq 1$".

In the case $m_{i+1}/m_i \geq a > 1$ this lemma is a classical result of Zygmund and its proof works word for word for the slightly more general case we need here.

To handle the case where N_o is neither very small nor very large (say $< N_e/4$) we need an easy lemma. We write F_e and F_o for the exponential sums corresponding to even and odd frequencies respectively and have

Lemma - "If $N_e - N_o > C \cdot N$ then $||F||_1 \geq ||F_o|| + (C/4)$".

If finally N_e and N_o are of comparable size, say $N_e/4 \leq N_o \leq N_e$ then we use the following inequality

$$||F||_1 \geq (1/2)(||Fe||_1 + ||F_o||_1) + \pi/(32||F||)$$

This inequality is a corollary of the following:

Lemma - Let E be a measurable subset of $[0,2\pi]$ with $|E| = \pi$ Then

$$\int_E |F| \geq 1/(4||F||_1)".$$

The last lemma holds also in the case of any trigonometric polynomial F of the form $F(x) = 1 + a_1\exp(ix) + a_2\exp(2ix) + \ldots$

However, it is not inconceivable that in the case of exponential sums the above inequality can be strengthened. If so, then we can have better results for the conjecture of Littlewood.

In particular, the method we sketched above will imply the conjecture of Littlewood if the following is true:

"There is an absolute positive constant C *such that if* $E \subset [0,2\pi]$ *and* $|E| = \pi$ *then* $\int_E |F| \geq C$*".*

The above statement is true in the extreme cases of frequencies in arithmetic progression or forming a lacunary sequence but it is apparently unknown if it is true in the general case.

We mention finally that if instead of exponential sums we consider polynomials F with coefficients C_k such that $|C_k| \geq 1$ the above proof yields $||F||_1 \geq C(\log||F||_2)^{1/2}$ and hence again $||F||_1 \geq C(\log N)^{1/2}$.

5. References: A detailed version of the content of 4 can be found in:

1) Pichorides S.K. "A lower bound for the L' norm of exponential sums I and II". Bulletin of the Greek Math. Society 18 (1977) and 19 (1978).

The work of Fournier is in preprint form now and will appear later. Some results mentioned in 3 and references to previous papers are in:

2) Roth F.K. On cosine polynomials corresponding to sets of integers. Acta Arithmetica XXIV (1973) 87-28.

SYMMETRY (OR SIMPLE MODULES) OF SOME BANACH ALGEBRAS

by Detlev Poguntke in Bielefeld

Recall that a Banach algebra A with isometric involution $a \to a^*$ is called symmetric if every element of the form a^*a, $a \in A$, has a real nonnegative spectrum. Several authors have investigated the question for which locally compact groups G the convolution algebra $L^1(G)$ is symmetric. Even for simply connected Lie groups G a necessary and sufficient criterion (e.g. in terms of the Lie algebra of G) is not known. Let $G = H \ltimes S$ be the Levi decomposition of the simply connected Lie group G with semisimple H and solvable S. Then the compactness of H is a necessary condition for the symmetry of $L^1(G)$ (non-compact semisimple Lie groups do never have symmetric group algebras, [1]). So, assyme that H is compact. Then the symmetry of $L^1(S)$ is sufficient for the symmetry of $L^1(G)$, [6]. $L^1(S)$ is symmetric if the Haar measure of S has polynomial growth, [7], which is equivalent, by [2], to the fact that all eigenvalues of all operators in the adjoint representation of S on the Lie algebra of S have absolute value 1. But there exist also other solvable Lie groups with symmetric group algebras, see e.g. [6]. On the other hand there is a lot of solvable Lie groups with nonsymmetric group algebras, [9]. Thus, for solvable Lie groups the question of symmetry seems to be very complicated.

In several cases, see [6] or [8], the question of symmetry can be reduced to the study of algebras of the following type which are the main theme in this article:

Let G be a locally compact group and let A be an involutive Banach algebra. Suppose that G acts strongly continuously on A, $(x,a) \to a^x$, by isometric $*$- isomorphisms. Then one can form the algebra $B = L^1(G,A)$ of left Haar integrable A-valued functions on G, see e.g. [3], with multiplication

$$(f * g)\ (x) = \int_G f(xy)^{y^{-1}}\ g(y^{-1})\ dy$$

and involution $f^*(x) = \Delta(x)^{-1} f(x^{-1})^{*x}$ where Δ denotes the modular function of G. Suppose further that U is a semisimple regular symmetric commutative Banach algebra on which G acts strongly continuously by isometric $*$-isomorphism, also denoted by $(x,u) \to u^x$. Then G acts also on the Gelfand space $\hat{U}$ of U. Fix any $\chi \in \hat{U}$. For $t \in G$ define $t\chi \in \hat{U}$ by $(t\chi)(u) = \chi(u^t) =: \hat{u}(t)$. Suppose that $t \to t\chi$ is an homeomorphism from G onto $\hat{U}$ (consequently, one can consider U as an algebra of functions on G via the Gelfand transform) and that the following two conditions hold:

(i) $U_o := \{u \in U | \hat{u}$ has compact support$\}$ is dense in U.

(ii) For every neighborhood W of e in G there exists $u \in U$, $u \neq 0$, and a continuous map $f: G \to U$ such that $\hat{u}$ is supported by W and $f(z)(x) = \hat{u}(xz)$ for all $x,z \in G$.

From these assumptions one can deduce, see [6], Theorem 4, that $L^1(G,U)$ is simple and symmetric and contains a lot of hermitian rank one projections (in fact, they span a dense two-sided ideal) which will be crucial in the sequel. Moreover, U and A are connected by the following assumptions: A has a U-module structure which is compatible with all the other operations,

i.e. $\quad |u a| \leq |u| \cdot |a|, \quad u(ab) = (ua)b = a(ub),$

$$(ua)^* = u^* a^*, \quad (ua)^x = u^x a^x$$

for all $a,b \in A$, $u \in U$, $x \in G$. In other words, when we form the Banach $*$-algebra $U \oplus A$ with the obvious operations then G acts strongly continuously by isometric $*$-isomorphisms on $U \oplus A$ and U is central in $U \oplus A$. Assume (and this is the last assumption) that UA is dense in A. What we want to do is to deduce properties of $B = L^1(G,A)$ from properties of A and vice versa. It is known that symmetry of A implies symmetry of B, [8]. In this paper I will give a different proof by a more general approach.

But before doing so, we should give an example of the situation described above in order to show that such a situation occurs "in nature".

Suppose that H is a Lie group with a normal subgroup $K = \mathbb{R}^2$ on which H acts (by inner automorphisms) via $\left\{\begin{pmatrix} 1 & 0 \\ x & 1 \end{pmatrix} \middle| \, x \in \mathbb{R}\right\}$. Then $Z = \{0\} \times \mathbb{R}$ is central in H. Let N be the centralizer of K. Then H is isomorphic to $\mathbb{R} \ltimes N$ and $L^1(H)$ is isomorphic to $L^1(\mathbb{R}, L^1(N))$ where the action of $\mathbb{R}$ is induced by the inner automorphisms. Choose a non-trivial character $\eta : Z \to T$ and form the algebra $L^1(H)_\eta$ of all measurable functions $f : H \to \mathbb{C}$ with $f(xz) = \eta(z) f(x)$ for (almost) all $z \in Z$, $x \in H$ and $\int_{H/Z} |f| < \infty$. Similarly, form $A := L^1(N)_\eta$ and $U := L^1(K)_\eta \simeq L^1(\mathbb{R})$. Then the triple $G = \mathbb{R}$, A, U satisfies the assumptions described above and $L^1(\mathbb{R}, A)$ is isomorphic to $L^1(H)_\eta$.

To attack the symmetry of $B = L^1(G,A)$ (in the general situation) we use the following characterization of symmetric algebras which is proved in Naimarks book "Normed rings" (in a different formulation), see also [5]:

An involutive Banach algebra C is symmetric iff for every algebraically irreducible representation ρ of C in a Banach space E (in the sequel, we will use the term "simple C-module" instead of "algebraically irreducible...") there exist a (topologically irreducible) $*$-representation π of C in the Hilbert space $\mathcal{H}$ and a non-zero interwining operator E in $\mathcal{H}$, i.e. $\mathrm{Hom}_C(E, \mathcal{H}) \neq 0$.

Thus, to decide whether a given involutive Banach algebra C is symmetric or not one may proceed in the following manner:

1° "Describe" all topologically irreducible $*$-representations of C.

2° "Describe" all simple C-modules.

3° Decide whether there exist interwining operators.

Of course, 1° and 2° are of independent interest, and for group algebras (or "related" algebras) a lot is known concerning 1°, but there seems to be more or less no information available on point 2° (except for group algebras of groups with "large" compact subgroups as semisimple Lie groups or semidirect products of compact groups with normal abelian subgroups, so-called motion-groups).

Another consequence of this characterization of symmetric algebras is the following: Let H be a Lie group as in the example. Then $L^1(H)$ is symmetric iff $L^1(H/Z)$ and $L^1(H)_\eta$ are symmetric for all non-trivial characters $\eta: Z \to T$. Therefore, one has to study the algebra $L^1(H)$ which are of the type discussed in this article.

From now on, we assume that G, U, A have the properties described above. For $B = L^1(G,A)$, we want to carry out the program $1^o - 3^o$. In fact, we can solve 1° more or less completely (Theorem 1) and 2° to such an extent (Theorem 2) that we can show that symmetry of A implies symmetry of B (Corollary to Theorem 2).

Fix $\chi \in \hat{U}$ once and for ever and define (for $t \in G$, $w \in U$)

$$\hat{w}(t) := \chi(w^t).$$

Then we have the formula $\hat{w^s}(t) = \hat{w}(st)$. Choose an $u \in U$, $0 \neq u = u^*$ such that $\hat{u}$ has a compact support. Form

$$v := \int_G u^y u^y u \, dy \in U .$$

v has the property that $\hat{v}(x) = \Delta(x)^{-1} \hat{u}(x) \int_G \hat{u}(z)^2 \, dz$ for all $x \in G$. Assume that $\int_G \hat{u}(z)^2 \, dz = \int_G \hat{v}(z)^2 \, dz$

(if we start with an arbitrary u then a certain constant multiple satisfies this equation).

Define $p : G \to U$ by $p(x) = v^x u$. p has the following properties:

(1) p is a continuous function with compact support, especially $p \in L^1(G,U)$.

(2) $p(x) = \Delta(x)^{-1} u^x v$ for all $x \in G$.

(3) $p = p^*$

(4) $p * p = p$

(5) $p * L^1(G,U) * p = \mathbb{C}\, p \neq 0$.

Let I be the closure of $\mathrm{Kern}\,\chi A$ in A, I is an $*$-ideal in A. Let $A' : A/I$ be the quotient algebra and denote by

$$Q : A \to A'$$

the quotient morphism.

From the assumptions it follows that $B = L^1(G,A)$ is an $L^1(G,U)$-bimodule and $L^1(G,U) * B * L^1(G,U)$ is dense in B. Especially, we can form $p * B * p$ which is a closed subalgebra of B. The first step is to establish a dense $*$-morphism $T : p * B * p \to A'$.

Let f be an element of $p * B$ ($\supset p * B * p$), i.e. $f = p * f$. Then we have

$$f(x) = (p * f)(x) = \int_G p(xy)^{y^{-1}} f(y^{-1})\, dy =$$

$$= \int_G \{\Delta(xy)^{-1} u^{xy} v\}^{y^{-1}} f(y^{-1})\, dy =$$

$$= \Delta(x)^{-1} \int_G \Delta(y)^{-1} u^x v^{y^{-1}} f(y^{-1})\, dy =$$

$$= \Delta(x)^{-1} \int u^x v^y f(y)\, dy = \Delta(x)^{-1} u^x \varphi$$

with $\varphi = \int_G v^y f(y)\, dy$.

Now, let $f \in p * B * p$, let φ be as above and define

$$T : p * B * p \to A'$$

by

$$Tf := ||\hat{u}||_2^{-2} \int_G Q(\hat{v}(t)\, \Phi^t)\, dt.$$

<u>Proposition</u> T is a dense $*$-morphism.

<u>Proof</u>. A straightforward computation shows that T is multiplicative

and involutive. To prove the density we observe the following two facts:

$$Q(wa) = \hat{w}(e)\ Q(a) \quad \text{for all} \quad w \in U, \quad a \in A.$$

If $g \in p*B$, $g(x) = \Delta(x)^{-1}\ u^x \varphi$,

then $T(g*p) = ||\hat{u}||_2^{-2} \int_G Q(\hat{v}(t)\varphi^t)\,dt.$

Without loss of generality we may assume that $\hat{v}(e) \neq 0$. To approximate a given $Q(a) \in A'$, for every neighborhood W of e we choose $w \in U$ such that $\hat{w}$ is supported by W, $\hat{v}\ \hat{w}$ is nonnegative and the integral over $||\hat{u}||_2^{-2}\ \hat{v}\ \hat{w}$ is one. Define $g \in p*B$ by $g(x) = \Delta(x)^{-1}\ u^x\ wa$. Then

$$T(g\ p) = ||\hat{u}||_2^{-2} \int_G Q(\hat{v}(t)\ w^t\ a^t)\ dt =$$

$$= ||\hat{u}||_2^{-2} \int_G \hat{v}(t)\ \hat{w}(t)\ Q(a^t)\ dt.$$

This integral is arbitrarily close to $Q(a) =$

$$= ||\hat{u}||_2^{-2} \int_G \hat{v}(t)\ \hat{w}(t)\ Q(a)\ dt \quad \text{if} \quad W \text{ is small enough.}$$

For a Banach $*$-algebra C, we denote by $\mathrm{Rep}(C)$ the equivalence classes of non degenerated $*$-representations in Hilbert spaces. The morphism T induces a map $\overline{T} : \mathrm{Rep}(A') \to \mathrm{Rep}(p*B*p)$. By restriction we get a map $S : \mathrm{Rep}(B) \to \mathrm{Rep}\ (p*B*p)$; if $\pi \in \mathrm{Rep}(B)$ is a representation in H then $S(\pi)$ is a representation in $\pi(p)\ H$; $\pi(p)$ is not zero because $B*p*B$ is dense in B. Moreover, we can construct a map $\mathrm{ind} : \mathrm{Rep}(A') \to \mathrm{Rep}(B)$ in the following way: Let ρ be a representation of A' in K. Let the representation $\tilde{\rho}$ of A in $H = L^2(G,K)$ be defined by $(\tilde{\rho}(a)\xi)(x) = \rho(Q(a^x))(\xi(x))$. For $t \in G$ and $\xi \in H$ let $\xi^t \in H$ be the function $\xi^t(x) = \xi(tx)$. Then we define the representation

$\pi = \mathrm{ind}(\rho)$ in H by

$$\pi(f)\xi = \int \tilde{\rho}(f(t)^{t^{-1}})\xi^{t^{-1}}\,dt.$$

Theorem 1 The diagram

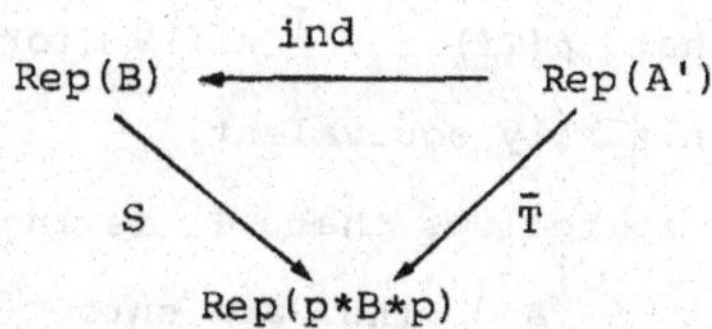

is commutative, and all three maps are bijections.

Remark Since all three maps are compatible with direct sums, irreducible representations correspond to irreducible representations.

Proof. For simplicity we assume that U is contained in A because the theorem is easily deduced from the corresponding theorem for $\tilde{A} = A \oplus U$, $\tilde{B} = L^1(G,\tilde{A})$, $\tilde{A}' = A' \oplus \mathbb{C}\,1$ and $p*\tilde{B}*p = p*B*p \oplus \mathbb{C}\,p$.

$\bar{T} = S \circ \mathrm{ind}$:

Let ρ be a $*$-representation of A' in K, let $H = L^2(G,K)$, $\tilde{\rho}$ and $\pi = \mathrm{ind}\,\rho$ be as above.

Then for $f \in B$, $\xi \in H$ and $s \in G$ we have

$$(\pi(f)\xi)(s) = \int_G dr\, \rho\,(Q(f(sr)^{r^{-1}}))\,(\xi(r^{-1})).$$

For $f \in p*B$, $f(x) = \Delta(x)^{-1}\Delta u^x\,\varphi$, we find

$$(\pi(f)\zeta)\;(s) = \Delta(s)^{-1}\int_G dr\;\hat{u}(s)\;\rho(Q(\varphi^r))\xi(r) =$$

$$= \|\hat{u}\|_2^{-2}\;\hat{v}(s)\int_G \rho(Q(\varphi^r))\xi(r)\;dr.$$

Especially, for $\varphi = v$, i.e. $f = p$, we get

$$(\pi(p)\xi)(s) = ||\hat{u}||_2^{-2}\ \hat{v}(s) \int_G \hat{v}(r)\xi(r)\,dr.$$

Therefore, $V : K \to H$ defined by

$(V\eta)(x) = ||\hat{v}||_2^{-1}\ \hat{v}(x)\eta$ is an isometry from K onto $\pi(p)\ H$.

A trivial computation shows that $\rho(Tf) = V^{-1}\ \pi(f)V$ for $f \in p*B*p$. Hence, $\rho \circ T$ and $S(\pi)$ are unitarily equivalent.

From the density of T it follows that $\bar{T}$ is injective. Therefore, it suffices to show that S and ind are onto.

<u>S is onto</u>: Let ρ be a $*$-representation of $p*B*p$. Since S is compatible with direct sums we may assume that ρ is cyclic, let ξ be a cyclic vector of norm 1 and let $f(x) := \langle\rho(x)\xi,\xi\rangle$ be the associated positive form. Define $F : B \to \mathbb{C}$ by $F(x) = f(p*x*p)$. As the computations in the proof of the Lemma in [8] show (approximate identities are not needed at this point), F is a positive form. Here we need that $\sum_{q,r\in P} q*B*r$ is dense in B (P = projections of rank one in $L^1(G,U)$) which follows from the fact that the linear span of P is dense in $L^1(G,U)$. Moreover, F can be extended as a positive form to $B \oplus \mathbb{C}\,1$, $\tilde{F} : B \oplus \mathbb{C}\,1 \to \mathbb{C}$ is defined by $\tilde{F}(x + \lambda) = F(x) + \lambda =$ $F(x) + \lambda = F(x) + \lambda f(p) = F(x + \lambda p)$. $\tilde{F}$ is positive because $\tilde{F}((x+\lambda)^*(x+\lambda)$ $\tilde{F}(p(x+\lambda)^*(x+\lambda)p) = F(|(x+\lambda)p|^*|(x+\lambda)p|) \geq 0$. From the GNS-construction it follows very easily that the representation π associated to F has the property that $S(\pi)$ is unitarily equivalent to ρ.

<u>ind is onto</u>: This fact can be deduced from a suitable version of the imprimitivity theorem, see e.g. [4]. But it can also be shown in the following simple way:

Let $C_\infty(G,A')$ be the Banach $*$-algebra of all continuous functions $G \to A'$ vanishing at infinity with the uniform norm and the pointwise operations. Let $F : A \to C_\infty(G,A')$ be defined by $(Fa)(t) = Q(a^t)$, F

is a dense $*$-morphism. $\mathcal{F}$ is G-equivariant if we let G act on $C_\infty(G, A')$ by left translations. Moreover, let $C^*(A')$ be the C^*-hull of A'. By composition, we find a G-equivariant dense $*$-morphism $\tilde{\mathcal{F}} : A \to C_\infty(G, C^*(A')) =: \tilde{A}$. Since every irreducible $*$-representation of A (which is a character on U) factorizes through $\tilde{\mathcal{F}}$, $\tilde{A}$ is just the C^*-hull of A.

Every $*$-representation of B factorizes through $B \to \tilde{B} = L^1(G, \tilde{A})$. Therefore, we may assume that $A = C_\infty(G,D)$ for some unital C^*-algebra D, $U = C_\infty(G)$ and G acts by left translations on A. In this case, A' is canonically isomorphic to D, and $T : p * B * p \to A'$ is an isometry onto A'. Now, let π be a $*$-representation of B, form the representation $S(\pi)$ of $p * B * p$ and define the representation ρ of A' by $\rho = S(\pi) \circ T^{-1}$. It is easy to see that $\mathrm{ind}(\rho)$ is unitarily equivalent to the given π.

Theorem 2: Let G, U, A and $B = L^1(G, A)$ be as always. Let p, A' and $T : p * B * p \to A'$ be as constructed above. If we associate to the simple B-module E the $p*B*p$ - module pE we get a bijection from the set of isomorphism classes of simple B-modules onto the set of isomorphism classes of simple $p * B * p$ - modules. Moreover, for every simple $p * B * p$ - module M there exists a simple A'-module M' such that $\mathrm{Hom}_T(M, M') \neq 0$, i.e. there is a non-zero linear map $R : M \to M'$ with $R(fm) = T(f)\, R(m)$ for $f \in p * B * P$, $m \in M$.

Corollary 1: Symmetry of A implies symmetry of B. To give at least one application, we formulate and prove (see also [6]).

Corollary 2: Let K be a compact group and let D be a symmetric Banach $*$-algebra on which K acts strongly continuously by isometric $*$-isomorphisms. Then $L = L^1(K, D)$ is symmetric.

Proof: As in the proof of Theorem 1 we may assume that U is contained in A.

If E is a simple B-module then $pE \neq 0$ because $B * p * B$ is dense in B, and it is easy to see that pE is a simple $p * B * p$ - module.

In order to show that $E \mapsto pE$ is a bijection we construct the inverse map. Let M be a simple $p*B*p$ - module. Then we form the induced B-module $M^i = B \otimes_{p*B*p} M$ and define $\tilde{M} := M^i/(\{\xi \varepsilon M^i \mid B\xi = 0\}$. M^i is not zero because the map $B \otimes M \to M$, $b \otimes \xi \to pbp\, \xi$, factorizes through M^i. pM^i (and hence $p\tilde{M}$) is isomorphic to the given M. Moreover, M^i (and hence $\tilde{M}$) is cyclic: for every ξ, $0 \neq \xi \varepsilon M$, $p \otimes \xi$ is a generator of M^i. Therefore, we can realize M^i as a quotient of B and we can introduce a Banach space structure on M^i and on $\tilde{M}$. So, $\tilde{M}$ is

(1) a cyclic Banach B-module, generated by every non-zero element in $p\tilde{M}$ with (2) $p\tilde{M} \cong M$.

From $(UA)^- = A$ it follows that B^2 is dense in B. But then $\tilde{M}$ has the property that

(3) $\xi \varepsilon \tilde{M}$, $B\xi = 0$ implies $\xi = 0$.

From (1), (2) and (3) it follows very easily that $\tilde{M}$ is a simple B-module.

The maps $E \mapsto pE$ and $M: \mapsto \tilde{M}$ are inverse to each other.

Now, let M be a simple $p*B*p$ - module, $M = pE$ for a (unique) simple B-module E. We want to construct a simple A'-module M' as in the theorem. The B-module structure on E is given by an A-module structure on E, $(a, \varepsilon) \mapsto a\varepsilon$, and a compatible G-action on E, $(t,\varepsilon) \mapsto \varepsilon^t$, compatible means that $(a\varepsilon)^t = a^t \varepsilon^t$. For $f \varepsilon B$ and $\varepsilon \varepsilon E$ we have the formula

$$f\varepsilon = \int_G (f(t)\varepsilon)^t \, dt = \int_G f(t)^{t^{-1}} \varepsilon^{t^{-1}} dt.$$

The functions $f \varepsilon B*p$ are of the form $f(x) = \psi^x u$ with some $\psi \varepsilon A$. If we define $A^\infty = \{\psi \varepsilon A \mid \int_G |\psi^x u| dx < \infty\}$ we get a surjective A-linear map $\Psi : A^\infty \to B*p$ (A^∞ and $B*p$ are considered as left A-modules); by the way, A^∞ is a two-sided, G-invariant, dense ideal in A.

For a fixed non-zero ξ in M we get by multiplication a B-linear map from $B * p$ onto E. Composition with Ψ gives a surjective A-linear map.

$$\Phi : A^{\infty} \to E$$

Especially, the kernel Ω of Φ is a left ideal in A.
For $\psi \in A^{\infty}$ we have the formula

$$\Phi(\psi) = \int_G \psi u^{t^{-1}} \xi^{t^{-1}} dt.$$

From $\xi = p\xi$ we obtain

$$\xi = \int_G p(t)^{t^{-1}} \xi^{t^{-1}} dt = \int_G v u^{t^{-1}} \xi^{t^{-1}} dt = \Phi(v).$$

Because Φ is G-equivariant, Ω is a G-invariant left ideal in A. Ω is contained in the annihilator $\mathrm{Ann}_A(\xi)$ of ξ in A because $\psi \in \mathrm{Kern}\Phi$ implies

$$0 = \int_G \psi u^{t^{-1}} \xi^{t^{-1}} dt, \text{ hence}$$

$$0 = \int_G v\psi u^{t^{-1}} \xi^{t^{-1}} dt = \int_G \psi v u^{t^{-1}} \xi^{t^{-1}} dt = \psi\xi.$$

Moreover, the integral representation of ξ shows that if we take $w \in U$ with $wv = v$ (clearly, such w's exist since $\hat{v}$ is compactly supported) then w is a right unit for the left ideal $\mathrm{Ann}_A(\xi)$. Therefore, $\mathrm{Ann}_A(\xi)$ (and hence Ω) is contained in a maximal modular left ideal Λ of A. From Schur's Lemma, it follows that U acts by a character on the simple A-module A/Λ, i.e. there exists $s \in G$ such that the kernel of $s\chi$ (recall $s\chi(w) = \chi(w^s)$ for $w \in U$) is contained in Λ. Λ^s is also a maximal modular left ideal in A. $M' := A/\Lambda^s$ is a simple A-module and in fact an A'-module because $\mathrm{Kern}\ \chi$ is contained in Λ^s. Since Ω is

G-invariant, Ω is contained in Λ^s.

To finish the proof of the theorem we have to construct a non-zero T-linear map $M \to M'$. Denote by $Q' : A \to M' = A/\Lambda^s$ the quotient map (Q' factorizes through $Q : A \to A'$) and define $\tilde{R} : A \to M'$ by

$$\tilde{R}(a) = \int_G \hat{v}(t)\, Q'(a^t)\, dt = \int_G Q'(v^t a^t)\, dt.$$

The restriction of $\tilde{R}$ to A^∞ factorizes through the quotient map $A^\infty \to A^\infty/\Omega$ and gives a linear map $A^\infty/\Omega \to M'$. Since Φ induces an isomorphism from A^∞/Ω onto E, we get a linear map $R : E \to M'$. We claim that the restriction of R to $M = pE$ is a dense T-linear map from M into M'.

Let $f, g \in p * B * p \subset p * B$, $f(x) = \Delta(x)^{-1}\, u^x\varphi$, $g(x) = \Delta(x)^{-1}\, u^x\psi$, and let $\varepsilon = g\,\xi \in M$. We have to show that $R(f\varepsilon) = Tf\, R\varepsilon$.

From $\xi = \int_G v u^{x^{-1}} \xi^{x^{-1}}\, dx$, it follows that

$$\varepsilon = g\xi = \int_G g(t)^{t^{-1}} \xi^{t^{-1}}\, dt = \int_G\int_G u\psi^t v^t u^{x^{-1}} \xi^{x^{-1}}\, dt\, dx = \Phi\Big(\int_G u\psi^t v^t\, dt\Big)$$

and, by definition, that $R\varepsilon = \tilde{R}\big(\int_G u v^t \psi^t\, dt\big) = \int_G \hat{v}(x)\, Q'\big(\int_G u^x v^{tx} \psi^{tx}\, dt\big)\, dx =$

$$= \int_G\int_G \hat{v}(x)\, \hat{u}(x)\, \Delta(x)^{-1} Q'(v^y \psi^y)\, dx\, dy = \int_G \hat{v}(y)\, Q'(\psi^y)\, dy, \text{ because}$$

$$\int_G \hat{v}(x)\, \hat{u}(x)\, \Delta(x)^{-1}\, dx = 1.$$

This is just the image of Tg under the quotient map $A' \to M'$. Since T is dense, R is dense, too.

Now, we compute $R(f\varepsilon)$:

$$f\varepsilon = \int_G f(y)^{y^{-1}} \varepsilon^{y^{-1}}\, dy = \int_G \Delta(y)^{-1} u\varphi^{y^{-1}} \varepsilon^{y^{-1}}\, dy = \int_G u\varphi^y \varepsilon^y\, dy =$$

$$= \int_G\int_G\int_G u\varphi^y u^y \psi^{ty} v^{ty} u^{x^{-1}} \xi^{x^{-1}}\, dt\, dx\, dy = \Phi\Big(\int_G\int_G \Delta(y)^{-1} u\varphi^y u^y \psi^z v^z\, dy\, dz\Big),$$

hence $R(f\varepsilon) = \int_G \hat{v}(t)Q'(\iint_{GG}\Delta(y)^{-1}u^t{}_\varphi{}^{yt}{}_u{}^{yt}{}_\psi{}^{zt}{}_v{}^{zt}\, dy\, dz)dt =$

$$= \int_G \hat{v}(t)Q'(\iint_{GG} \Delta(y)^{-1}\Delta(t)\,u^t f^y u^y{}_\psi{}^z v^z \Delta(t)^{-2}\, dy\, dz)\, dt =$$

$$= \iiint_{GGG}\Delta(y)^{-1}\Delta(t)^{-1}\hat{v}(t)\hat{u}(t)\hat{v}(z)\hat{u}(y)Q'(\varphi^y\psi^z)\, dt\, dy\, dz =$$

$$= \iint_{GG}\Delta(y)^{-1}\hat{u}(y)\hat{v}(z)Q'(\varphi^y\psi^z) = ||\hat{u}||_2^{-2}\iint_{GG}\hat{v}(y)\hat{v}(z)Q'(\varphi^y\psi^z)\,dy\, dz\, .$$

From the definition of Tf and the formula for $R\varepsilon$ it follows that $R(f\varepsilon) = Tf\, R\varepsilon$.

The theorem is proved. For the proof of corollary 1 we use (of course) the characterization of symmetry by simple modules. Let E be a simple B-module, let $M = pE$. We may assume that $E \doteq \tilde{M} = M^i/\{\gamma \in M^i | B\gamma = 0\}$. From the theorem it follows that there exists a non-zero T-linear map from M into a simple A'-module M' . By assumption we find a non-degenerated (irreducible)$*$-representation ρ of A' in the Hilbert space K and a non-zero A'-interwinning operator $M' \to K$. By composition we get a non-zero T-linear map $R : M \to K$. Let $\pi = \mathrm{ind}(\rho)$ be the induced representation of B in $H = L^2(G,K)$. We claim that we can embed E into H.

Define $\tilde{R} : M \to H$ by $(\tilde{R}\xi)(t) = \hat{v}(t)R\xi$ and $V : B \otimes M \to H$ by $V(g \otimes \xi) = \pi(g)(\tilde{R}\xi)$.

From the fact that $R : M \to K$ is T-linear one easily deduces the formula

$$\pi(f)(\tilde{R}\xi) = \tilde{R}(f\xi)$$

for $f \in p * B * p$ and $\xi \in M$.

Therefore, the map $V : B \otimes M \to H$ factorizes through $M^i = B \otimes_{p*b*p} M$ and gives a B-linear map $\tilde{V} : M^i \to H$. Since π is a non-degenerated representation the subspace $\{\gamma \in M^i | B\gamma = 0$ is contained in the kernel of $\tilde{V}$, and finally we find a non-zero B-linear map from $\tilde{M} = E$ into H.

<u>Proof of Corollary 2</u>. Let $C = C(K,D)$ be the Banach *-algebra of all continuous functions from K into D with the pointwise operations and the uniform norm. We let K act on C by left translations, i.e. $f^t(x) = f(tx)$ for $t, x \in K$ and $f \in C$. The map $\alpha : D \to C$ defined by

$$\alpha(d)(x) = d^x$$

is a K-equivariant injective *-isomorphism from D onto a closed subalgebra of C and induces a *-isomorphism from $L = L^1(K,D)$ onto a closed subalgebra of $L^1(K, C)$. Since closed *-subalgebra of symmetric Banach algebras are symmetric it suffices to show that $L^1(K, C)$ is symmetric. But the triple $G = K$, $A = C$, $U = C(K, \mathbb{C})$ satisfies the usual assumptions of this paper and, consequently, the symmetry of C implies the symmetry of $L^1(K, C)$.

The selection of p was rather arbitrary. So, one could think that it might be useful to study different p's. That this is not the case is shown by the following remark which we state without proof.

<u>Remark</u> Let $q \in L^1(G, U)$ be another hermitean rank one projector. Then q has a similar structure as p even though the representing functions in U need not to have compactly supported Gelfand transforms. Nevertheless, in a similar way one can construct a *-morphism T_q from $q * B * q$ into A'. But there exists a unique (partial isometry) $k \in L^1(G, U)$ with $k^* * k = p$ and $k * k^* = q$ which gives rise to a *-isomorphism from $p * B * p$ onto $q * B * q$ such that the diagram

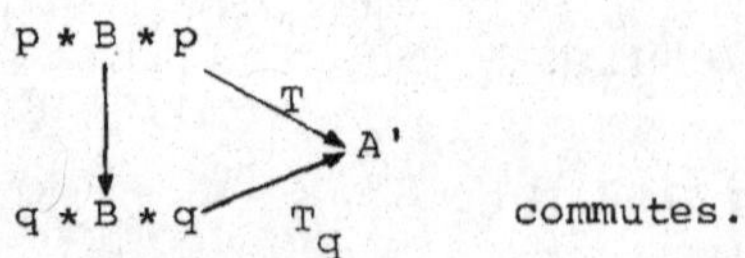

commutes.

We conclude this article with some open questions.

(1) Does the symmetry of B imply the symmetry of A ?
The answer is yes if A is commutative (since $T : p * B * p \to A'$ is

dense and $p * B * p$ is symmetric every multiplicative functional on A' is hermitean) or if G is discrete because in this case the map $T : p * B * p \to A'$ is onto. In general, T is not onto, there is an example in [9]. I don't know the answer even for compact abelian G.

(2) Let C be a symmetric Banach $*$-algebra. The characterization of symmetry used in this paper tells us that for every simple C-module E there exists an irreducible $*$-representation in H and a non-zero C-linear map $E \to H$. But is $E \to H$ unique in the sense that if there are two embeddings $E \to H_j$ (j = 1,2) that there exists an interwinning operator $H_1 \to H_2$ such that

commutes ?

The answer is yes if C is the group algebra of a connected nilpotent Lie group. Suppose that A has this uniqueness property. Is it true that $B = L^1(G, A)$ has this uniqueness property ?

(3) Let [NeS] be the class of all Banach $*$-algebras C with the property that every topologically completely irreducible Banach C-module is Naimark-equivalent to an irreducible $*$-representation of C. Of course, every algebra in this class is symmetric. Moreover, the group algebras of connected two step nilpotent Lie groups and of motion groups, see [10], are in [NeS]. Is it true that B is in [NeS] provided that A is in [NeS]?

(4) It is known that in a certain sense it is impossible to classify the irreducible $*$-representations of a non-type I group but possibly it is simpler to classify the simple modules over a symmetric group algebra.

To test this one should treat the following example:

Let $C(T^2)$ be the C*-algebra of all continuous functions on the torus, let $\mathbb{R}$ act on $C(T^2)$ by

$$f^t(z, w) = f(e^{it}z, e^{i\alpha t}w)$$

for some irrational α, and form the algebra $L = L^1(\mathbb{R}, C(T^2))$; this algebra is closely related to the group algebra of the Mautner group. If we let act L on $E = C(T^2)$ by

$$f\xi = \int_{\infty}^{\infty} f(t)^{t^{-1}} \xi^{t^{-1}} dt$$

where the product of $f(t)^{t^{-1}} \varepsilon C(T^2)$ and $\xi^{t^{-1}} \varepsilon C(T^2)$ is the usual pointwise product in $C(\ ^2)$, we get a simple L-module. From this L-module we can construct further simple L-modules if we apply first the automorphisms $U_z : L \to L$ $(z \varepsilon \mathbb{R})$ defined by

$$(U_z f)(t) = e^{itz} f(t) .$$

Are <u>all</u> simple L-modules obtained in this way ?

References

[1] J.W. Jenkins, Nonsymmetric group algebras, Studia Math. 45 (1973), 295-207.

[2] J.W. Jenkins, Growth of Connected Locally Compact Groups, J.Funct. Anal. 12 (1973), 113-127.

[3] H. Leptin, Verallgemeinerte L^1-Algebren und projektive Darstellungen lokalkompakter Gruppen, Inventiones math. 3 (1967), 257-281, 4 (1967), 68-86.

[4] H. Leptin, Darstellungen verallgemeinerter L^1-Algebren II in Lectures on Operator Algebras, Lecture Notes in Mathematics 247 (1972), 251-307.

[5] H. Leptin, Symmetrie in Banachschen Algebren, Arch. d. Math. 27 (1976), 394-400.

[6] H. Leptin und D. Poguntke, Symmetry and nonsymmetry for locally compact groups, to appear in J. Funct. Anal.

[7] J. Ludwig, A class of symmetric and a class of Wiener group algebras, to appear in J. Funct. Anal.

[8] D. Poguntke, Nilpotente Liesche Gruppen haben symmetrische Gruppenalgebren, Math. Ann. 227 (1977), 51-59.

[9] D. Poguntke, Nichtsymmetrische sechsdimensionale Liesche Gruppen, to appear in J. reine angew. Math.

[10] R. Gangolli, On the symmetry of L^1-algebras of locally compact motion groups and the Wiener Tauberian theorem. J. Funct. Anal. 25 (1977), 244-252.

Fakultät für Mathematik der Universität, D-4800 Bielefeld, Postfach 8640.

SOME PROBLEMS ON SPECTRAL SYNTHESIS

J.D. Stegeman

1. Introduction

Two famous unsolved problems in harmonic analysis are the union problem for sets of spectral synthesis, and the problem whether each set of spectral synthesis is a Calderón set (the C-set-S-set problem, in the terminology of [1]; Problem III in this note). Examining these problems we observe that they are really special cases of two very natural problems (Problems I and II). Some other problems are formulated (Problems IV, V, VI) and relations between the problems II to VI are analysed.

To do all this a new type of set is defined:

Ditkin set (an old name for a new notion). Some properties of these sets are also given.

Further a notation is introduced to facilitate handling of definitions and proofs where local arguments are involved. The paper also aims at making some propaganda for that notation, which may be useful also in other situations.

2. Preliminaries

Let G be a locally compact abelian group, $\hat{G}$ its dual group. We consider the Fourier algebra $A = A(G) = \mathcal{F}(L^1(\hat{G}))$ with norm $\|f\|_A = \|F\|_1$, if $f = \hat{F}$.

Let I be an ideal in A. The cospectrum of I is the closed form $\text{cosp}(I) = \{x \in G \mid f(x) = 0 \ (\forall f \in I)\}$.

An ideal of its closure have the same cospectrum.

For a closed set $E \subset G$ one considers the following ideals:

$I_E = \{f \in A \mid f|_E = 0\}$,

$K_E = \{f \in A \mid \mathrm{Supp}(f) \cap E = \emptyset\}$,

$L_E = \{f \in K_E \mid \mathrm{Supp}(f)$ is compact$\}$.

Then I_E is the largest, L_E the smallest ideal with cospectrum E. One has $\overline{L_E} = \overline{K_E}$, and this closure is called J_E. It is the smallest closed ideal with cospectrum E. The set E is a set of spectral synthesis (a <u>synthesis set</u>) if $J_E = I_E$, thus if each $f \in I_E$ can be approximated by functions $g \in K_E$. If such an approximation is possible with functions fh with $h \in K_E$, then E is called a <u>Calderón set</u> (or a <u>C-set</u>).

3. <u>Two problems</u>

(i) Let E_1 and E_2 be two closed subsets of G. Put $E_1 \cup E_2 = E$. Then, trivially, $I_E = I_{E_1} \cap I_{E_2}$ and $K_E = K_{E_1} \cap K_{E_2}$. But for J_E one has only $J_E \subset J_{E_1} \cap J_{E_2}$ (because $\overline{K_{E_1} \cap K_{E_2}} \subset \overline{K_{E_1}} \cap \overline{K_{E_2}}$). A natural question to ask is now:

<u>Problem I</u>. Does $J_E = J_{E_1} \cap J_{E_2}$ always hold ?

In the special case that E_1 and E_2 are synthesis sets one has $J_{E_1} \cap J_{E_2} = I_{E_1} \cap I_{E_2} = I_E$, and the problem reduces to: is $J_E = I_E$, i.e., is E a synthesis set ? This is the "classical" union problem for synthesis sets.

(ii) A closed set $E \subset G$ is a Calderón set if $f \in \overline{fK_E}$ for all $f \in I_E$ (cf. section 2). Now $\overline{fK_E} \subset \overline{K_E} = J_E$, therefore the only functions $f \in I_E$ that can at all belong to $\overline{fK_E}$ are those in J_E. It is natural to inquire $f \in \overline{fK_E}$ only for functions $f \in J_E$. We are led to the following:

<u>Definition</u>. A closed set E G is a <u>Ditkin set</u> (or a <u>D-set</u>) if

$f \in \overline{fK_E}$ for all $f \in J_E$.

A natural question to ask is:

Problem II. Are all closed sets Ditkin sets ?

A Calderón set is just a set which is both synthesis and Ditkin. Hence the classical problem whether all synthesis sets are Calderón sets is Problem II asked for synthesis sets only. Let us state this problem explicitly:

Problem III. Are all synthesis sets Calderón sets ?

Remark. Calderón sets are sometimes called Ditkin sets, for instance in [5]. But Calderón set or C-set now seems the most common, cf. [1], [2], [4], so that the name Ditkin set seemed to be available for the new notion. Moreover, this name happens to fit nicely with Reiter's terminology ([3]):

Calderón = Wiener - Ditkin = synthesis and Ditkin.

4. Localisation

Equality of ideals of A is a local matter, in the sense that two ideals are equal if they are equal locally at each point of G and at infinity. To make this more precise, and to make a good use of it, we now introduce some notation.

Let X be a topological space and $a \in X$ a point. We will write $=_a$, ε_a, $\subset_a$, ... to indicate validity of the relations $=$, ε, $\subset$, ... in a nhd (neighbourhood) of a. Thus, for functions f and g on X, $f =_a g$ means that $f|_U = g|_U$ for some nhd U of a; for subsets E and F of X, $E =_a F$ means that $E \cap U = F \cap U$ for some nhd U of a; for a collection I of functions on X, $f \varepsilon_a I$ means that $f =_a g$ for some $g \varepsilon I$. For I and J two collections of functions on X, $I \subset_a J$ will mean that $f \varepsilon_a J$ for all $f \in I$.

For a locally compact space X, $=_\infty, \varepsilon_\infty, \subset_\infty, \ldots$ will denote validity of $=, \varepsilon, \subset, \ldots$ outside some compact subset of X. We shall write $f \neq_a g$, $E \neq_a F$, $f \not\varepsilon_a I$, etc. if the relations $f =_a g$, $E =_a F$, $f \varepsilon_a I$, etc. do not hold.

It is trivial that sets $\{x \mid f =_x g$, $\{x \mid f \varepsilon_x I\}$, etc. are open, and hence $\{x \mid f \neq_x g\}$, $\{x \mid f \not\varepsilon_x I\}$, etc. are closed. But a set $\{x \mid I =_x J\}$ need not be open, because the nhds may depend on the functions. Example: $X = \mathbb{R}$, I the functions vanishing in a nhd of 0, J the zero function. Then $\{x \mid I =_x J\} = \{0\}$.

We give some examples to illustrate the use of the notation.

$E =_a \emptyset \Longleftrightarrow a \notin \bar{E}$ ($a \notin E$ if E is closed).

$E =_a X \Longleftrightarrow a \in \mathring{E}$ ($\mathring{E}$ is the interior of E).

$\emptyset \neq_a E \neq_a X \Longleftrightarrow a \in \partial E$ (∂E is the boundary of E).

$E =_a \{a\} \Longleftrightarrow$ a is an isolated point of E.

$\{x \mid f \neq_x 0 = \operatorname{Supp} f$.

$f =_\infty 0 \Longleftrightarrow f$ has compact support.

For the Fourier algebra A we have:

$\operatorname{cosp}(I) = \{x \in G \mid I \neq_x A\}$.

$K_E =_x A$ is $x \notin E$, $K_E =_x O$ if $x \in E$ (O stands for the zero ideal $\{0\} \subset A$). $L_E =_x K_E$ $(\forall x \varepsilon G)$, $L_E =_\infty O$.

We also have the well-known

Localisation Lemma. Let $I \subset A$ be an ideal. If a function $f\colon G \to \mathbb{C}$ satisfies $f \varepsilon_x I$ for all $x \in G \cup \{\infty\}$, then $f \in I$ (in particular $f \in A$). If I is a closed ideal, and $f \in A$, then already $f \varepsilon_x I$ $(\forall x \varepsilon G)$ implies $f \in I$.

As a corollary we obtain the precise form of the statement made at the beginning of this section:

<u>Corollary</u>. Let I and J be two ideals in A.
If $I =_x J$ for all $x \in G \cup \{\infty\}$ then $I = J$.
If I and J are closed ideals and $I =_x J$ for all $x \in G$, then $I = J$.

5. A fundamental lemma

The following simple lemma provides the crucial step in several proofs.

<u>Lemma</u>. Let $f \in A$ be a function, $I \subset A$ an ideal, $E \subset G$ a closed set.
If $f \in_x I$ for all $x \notin E$, then $fL_E \subset I$.
If moreover I is closed, then $\overline{fL_E} = \overline{fK_E} = \overline{fJ_E} \subset I$.

<u>Proof</u>. If $x \notin E$ then $fL_E =_x fA \subset I$.
If $x \in E \cup \{\infty\}$ then $fL_E =_x f0 = 0 \subset I$.
Hence $fL_E \subset I$ by the localisation lemma.
The second part follows from the fact that $\overline{L_E} = \overline{K_E}$ and that, quite generally, for $X \subset A$, $\overline{f\overline{X}} = \overline{fX}$, by continuity.

6. Points of non synthesis

Let $E \subset G$ be a closed set. We define

$$B_E = \{x \in G \mid I_E \neq_x J_E\} = \bigcup_{f \in I_E} \{x \in G \quad f \notin_x J_E\},$$

$$\Gamma_E = \bigcup_{f \in J_E} \{x \in G \quad f \notin_x \overline{fK_E}\}.$$

One has $B_E \subset \partial E$, $\Gamma_E \subset \partial E$. (Indeed, if $x \notin E$ then $J_E =_x I_E =_x A$, and if $x \in \mathring{E}$ then $J_E =_x I_E =_x 0$. Likewise for Γ_E.) E is a synthesis set iff $B_E = \emptyset$, a Ditkin set iff $\Gamma_E = \emptyset$.

<u>Remark</u>. B_E and Γ_E are unions of closed sets, but it seems to be unknown whether they are themselves closed in general. For <u>metrizable</u> G one can prove that B_E is closed.

Part (i) of the following theorem is well known, in various formulations.

Theorem 1. Let $E \subset G$ be a closed set.

(i) If there exists a Calderón set C such that $B_E \subset C \subset E$, then $B_E = \emptyset$, thus E is a synthesis set.

(ii) If there exists a Ditkin set D such that $\Gamma_E \subset D \subset E$, then $\Gamma_E = \emptyset$, thus E is a Ditkin set.

Proof. (i) Take $f \in I_E$. If $x \notin C$ then $x \notin B_E$, thus $f \in_x J_E$. Hence, by the lemma, $\overline{fK_C} \subset J_E$. Now, $C \subset E$, thus $I_C \supset I_E$, thus $f \in I_C$. Thus $f \in \overline{fK_C}$, because C is Calderón. Thus $f \in J_E$ and $\{x \mid f \notin_x J_E\} = \emptyset$. It follows that $B_E = \emptyset$.

(ii) Take $f \in J_E$. If $x \notin D$ then $x \notin \Gamma_E$, thus $f \in_x \overline{fK_E}$. Hence, by the lemma, $\overline{fK_D} \subset \overline{fK_E}$. Now $D \subset E$, thus $J_D \supset J_E$, thus $f \in J_D$. Thus $f \in \overline{fK_D}$, because D is Ditkin. Thus $f \in \overline{fK_E}$ and $\{x \mid f \notin_x \overline{fK_E}\} = \emptyset$. It follows that $\Gamma_E = \emptyset$.

Theorem 1 (i) has familiar consequences, like: the union of two synthesis sets is again a synthesis set, provided their intersection is a Calderón set; or provided one of them is a Calderón set.
For Ditkin sets one obtains results similar to well-known results for Calderón sets: the union of two Ditkin sets is again a Ditkin set; if ∂E is a Ditkin set, then E is a Ditkin set as well.

As an example, let us give a short proof of the following recent theorem of C.R. Warner ([5; theorem 4]):

Theorem 2. Let E_1 and E_2 be closed subsets of G.
If $E_1 \cap E_2$ is a Calderón set and $E_1 \cup E_2$ is a synthesis set, then E_1 and E_2 are synthesis sets.

Proof. Put $E_1 \cup E_2 = E$. If $x \notin E_1$, then $E =_x E_2$. hence
$I_{E_2} =_x I_E = I_E =_x J_{E_2}$, thus $x \notin B_{E_2}$.
If $x \notin E_2$ then a fortiori $x \notin B_{E_2}$. It follows that $B_{E_2} \subset E_1 \cap E_2$.

Hence $B_{E_2} = \emptyset$ by Theorem 1 (i), thus E_2 is a synthesis set. Hence so is E_1.

We see from theorem 1 (i) that $\overline{B_E}$ cannot be a Calderón set, unless it is empty. But can it be a synthesis set ? This seems to be unknown. We state explicitly:

Problem IV. If E is a non-synthesis set, does it follow that $\overline{B_E}$ is a non-synthesis set ?

If Problem III has a positive answer, then of course $\overline{B_E}$ has to be non-synthesis in order to be non-Calderón. Thus yes for Problem III implies yes for Problem IV.

7. Functions of non-synthesis

In this section we consider the capability of a function $f \in A$ to establish non-synthesis sets and non-Calderón sets. This will lead us to yet two other related problems.

For $f \in A$ put $Z_f = \{x \mid f(x) = 0\}$. Define mappings β_f and γ_f, with domain (and range) the closed subsets of Z_f, by

$$\beta_f(E) = \{x \in G \mid f \notin_x J_E\},$$

$$\gamma_f(E) = \{x \in G \mid f \notin_x \overline{fK_E}\}.$$

Henceforth we will mostly drop the subscript f.

The mappings β and γ are decreasing ($\beta(E) \subset E$, $\gamma(E) \subset E$) and monotonous ($E_1 \subset E_2$ implies $\beta(E_1) \subset \beta(E_2)$, $\gamma(E_1) \subset \gamma(E_2)$). We have further $\beta(E) \subset \gamma(E) \subset \partial E \cap \partial Z$.

If $S \subset Z$ is a synthesis set, then $\beta(E) = \emptyset$ for all $E \subset S$.

If $C \subset Z$ is a Calderón set, then $\beta(E) = \gamma(E) = \emptyset$ for all $E \subset C$.

We note further that $\beta(Z) = \gamma(Z)$, because $\overline{fK_Z} = J_Z$.

An application of Theorem 1 yields for a Calderón set $C \subset Z$ and a Ditkin set $D \subset Z$:

(a) $\beta(E) \subset C$ implies $\beta(E) = \emptyset$

(b) $\gamma(E) \subset D$ implies $\gamma(E) = \emptyset$.

In particular $\beta(E)$ and $\gamma(E)$ are never Calderón sets, unless they are empty. We remark yet that one cannot conclude (b) from (a), nor (a) from (b), by a mere application of the inclusion $\beta(E) \subset \gamma(E)$.

We prove the following:

<u>Theorem 3</u>. (a) $\gamma \circ \gamma = \gamma$

(b) $\gamma \circ \beta = \beta$

<u>Proof</u>. (a) If $x \notin \gamma(E)$ then $f \,\varepsilon_x\, \overline{fK_E}$. By the lemma (section 5) this implies $\overline{fK_{\gamma(E)}} \subset \overline{fK_E}$. The opposite inclusion follows from $\gamma(E) \subset E$, thus $\overline{fK_{\gamma(E)}} = \overline{fK_E}$. But then $\gamma(\gamma(E)) = \gamma(E)$.

(b) If $x \notin \beta(E)$ then $f \,\varepsilon_x\, J_E$. By the lemma this implies $\overline{fK_{\beta(E)}} \subset J_E$. Therefore $\{x \mid f \not\varepsilon_x\, \overline{fK_{\beta(E)}}\} \supset x \mid f \not\varepsilon_x\, J_E\}$, thus $\gamma(\beta(E)) \supset \beta(E)$. But γ is decreasing, thus in fact equality must hold.

We observe that one cannot prove $\beta \circ \beta = \beta$ in the same manner, because in general $J_{\beta(E)} \neq J_E$. Nevertheless β might be an idempotent mapping. It is even conceivable that in fact β and γ are the same mapping. We state explicitly:

<u>Problem V</u>. Is β_f idempotent for all $f \varepsilon A$?

<u>Problem VI</u>. Is $\beta_f = \gamma_f$ for all $f \varepsilon A$?

If the answer to the latter problem is <u>yes</u>, then the same is true, trivially, for the former.

We can further prove:

<u>Theorem 4</u>. (a) Problem VI is equivalent to Problem II. (b) A positive answer to Problem V implies a positive answer to Problem IV.

<u>Proof</u>. (a) If $\beta_f(E) \neq \gamma_f(E)$ for some set $E \subset Z_f$ and some $f \in I_E$, then take $a \in \gamma_f(E) \setminus \beta_f(E)$. Next take a function $\tau \in A$ with $\tau =_a 1$ and $\operatorname{supp} \tau \cap \beta_f(E) = \emptyset$ (such functions τ exist in A). Put $\tau f = g$. Then $g \in_x J_E$ for all $x \in G$, as is easily seen, thus $g \in J_E$, thus $\beta_g(E) = \emptyset$. But, for x in a nhd of a, $g =_x f$, thus $gK_E =_x fK_E$. Thus $\overline{gK_E} =_a \overline{fK_E}$ ($=_a$ is not stable under taking closures, but it is stable if $=_x$ holds in a nhd of a). Now $g =_a f$. Thus $g \notin_a \overline{gK_E}$, and $\gamma_g(E) \neq \emptyset$.

A set E is a Ditkin set iff $f \in J_E$ implies $f \in \overline{fK_E}$, thus iff $\beta_f(E) = \emptyset$ implies $\gamma_f(E) = \emptyset$ for all $f \in I_E$ (i.e. all f with $Z_f \supset E$). But, as we have shown above, this is equivalent to $\beta_f(E) = \gamma_f(E)$ for all $f \in I_E$. This proves (a).

(b) Suppose a non-synthesis set E exists such that $\overline{B_E}$ is a synthesis set. Take $f \in I_E \setminus J_E$. Then $\emptyset \neq \beta_f(E) \subset B_E$, therefore $\beta_f(\beta_f(E)) = \emptyset$ (because $\overline{B_E}$ synthesis), and thus β_f is not idempotent.

Summarizing we obtain the following implication scheme:

$$\begin{array}{lcl} \text{II : yes} \rightarrow \text{III : yes} & \searrow & \\ \updownarrow & & \text{IV : yes.} \\ \text{VI : yes} \rightarrow \text{V : yes} & \nearrow & \end{array}$$

So, formally, we can distinguish six types of locally compact abelian groups:

Type 1 : II : yes.

Type 2 : II : no ; III, V : yes.

Type 3 : III : no ; V : yes.

Type 4 : V : no : III : yes.

Type 5 : III, V : no; IV : yes.

Type 6 : IV : no.

Does any of the four implications in the opposite direction hold ? This would of course reduce the number of possible types. Discrete groups are of type 1. Does any of the other types of l.c.a. groups really occur ? How many types occur ? One guess would be that all l.c.a. groups are of type 1, another guess that all non-discrete l.c.a. groups are of type 6. At any rate, these questions and problems may add some perspective to the classical "C-set-S-set problem".

REFERENCES

[1]. J.J. Benedetto, Spectral synthesis, Teubner (1975).

[2]. E. Hewitt, K.A. Ross, Abstract Harmonic analysis, vol. I, II, Springer, Berlin (1963, 1970).

[3]. H. Reiter, Classical harmonic analysis and locally compact groups, Oxford (1968).

[4]. W. Rudin, Fourier analysis on groups, Interscience, New York (1962).

[5]. C.R. Warner, A class of spectral sets, Proc. Amer.Math.Soc. 57, 99-102 (1976).

Mathematisch Instituut
der Rijksuniversiteit
Utrecht
The Netherlands

ON SPECTRAL ANALYSIS IN LOCALLY COMPACT GROUPS

by Yitzhak Weit

Introduction

Wiener's theorem in spectral analysis of bounded functions on the real line, states that every non-zero translation - invariant, w*-closed, subspace of $L_\infty(\mathbb{R})$ contains an exponential function. For a non-abelian, locally compact group G, the two-sided analogue of Wiener's theorem may be formulated as follows: Every non-zero, two-sided invariant, w*-closed subspace of $L_\infty(G)$ contains an indecomposable positive definite function. It was proved in [7] that the two-sided analgue of Wiener's theorem hold for all connected nilpotent Lie groups, and for all semi-direct products of abelian groups.

We are going to study the problems of one-sided spectral analysis of bounded functions on locally compact groups.

For a non-abelian group G, we may pose the following problems:

(i) Find a minimal class F, $F \subset L_\infty(G)$, such that every right invariant, w*-closed subspace of $L_\infty(G)$ contains a member of F.

(ii) Determine the fundamental functions of F, i.e. the functions which generate minimal, non-zero, right invariant, w*-closed subspace.

(iii) Is every non-zero, right invariant, w*-closed subspace of $L_\infty(G)$ contains a fundamental function of F ?

For the group G_L of the linear transformations on the real line of the form $ax+b$, $a>0$, $b \in \mathbb{R}$, the problems of two-sided spectral analysis, are easily solved, as seen by the following:

Let M denote the w*-closed subspace spanned by the two-sided translates of a non-zero function f, $f \in L_\infty(G_L)$. The subspace M contains all the functions of the form $f(\lambda a, \mu b+\beta a+\alpha)$ for every $\lambda>0$, $\mu>0$, $\alpha \in \mathbb{R}$, $\beta \in \mathbb{R}$. Let $\mu=1$ and $\beta=0$. Then, applying Wiener's theorem for $\mathbb{R}\times\mathbb{R}^+$ ($\mathbb{R}^+$ denotes the multiplicative group of positive reals) we deduce that $a^{i\theta}e^{i\tau b} \in M$ for some $\theta \in \mathbb{R}$ and $\tau \in \mathbb{R}$.

For a sequence $\mu_n \to 0$ we have

$$a^{i\theta}e^{i\tau\mu_n b} \underset{n}{\overset{w}{\to}} a^{i\theta}.$$

Thus, every two-sided invariant, w*-closed subspace of $L_\infty(G_L)$ contains a character of G_L [8].

However, the situation is completely different for the one-sided problem: the subspace $\{\phi(a)e^{ib}: \phi \in L(\mathbb{R})\}$ is a minimal, right invariant, w*-closed subspace of $L_\infty(G_L)$, which does not contain any indecomposable positive definite function.

To generalize the notion of an exponential function to an arbitrary group G, we follow [4]. Let X be an homogeneous space of G. A bounded measurable function $\sigma(g,x)$ on $G \times X$ which satisfies

(1) $\sigma(g\gamma,\xi) = \sigma(g,\gamma\xi)\sigma(\gamma,\xi)$ for every

$g \in G$, $\gamma \in G$ and $\xi \in X$ is called a multiplier.

The functions $\sigma(\cdot,\xi)$ are called multiplier functions. Notice, that if X consists of a single point, then (1) defines a character on G. Thus the bounded exponentials are the multiplier functions on $\mathbb{R}$ corresponding to a trivial X.

In [4], a theorem of Choquet and Deny [1] for $\mathbb{R}^n$ is generalized for semi-simple Lie groups. The positive exponentials in the integral representation are replaced by multiplier functions with respect to the homogeneous space B(G), where B(G) is a boundary of G. The Poisson formula for harmonic functions given in [3] implies that every harmonic function in G is an integral over multiplier functions with respect to B(G). In [2], a similar Poisson formula was obtained for the group G_L with respect to the homogeneous space $\mathbb{R}$ as a boundary of the group. These results suggest that the multiplier functions of G_L, with respect to B(G). In [2], a similar Poisson formula was obtained for the group G_L with respect to the homogeneous space $\mathbb{R}$ as a boundary of the group. These results suggest that the multiplier functions of G_L, with respect to $\mathbb{R}$, may play the role of the class F for one-sided spectral analysis. However, it was proved in [9], that there exists a right invariant, non-zero, w*-closed subspace of $L_\infty(G_L)$ which does not contain any multiplier function with respect to $\mathbb{R}$.

In this work we investigate the problems of one-sided spectral analysis for the two-dimensional Euclidean motion group, M(2). In section 1, the first two problems, stated above, are solved where the class F turns to be the multiplier functions of M(2) with respect to $\mathbb{R}^2$. In section 2, the third question is answered in the negative.

That is, there exists a non-zero right invariant, w*-closed subspace of $L_\infty(M(2))$ which does not contain a minimal, right invariant, non-trivial, w*-closed subspace.

Section 1

Let $M(2) = \{(e^{i\alpha},z) : \alpha \in \mathbb{R},\ z \in \mathbb{C}\}$ denote the two dimensional motion group. For $R \geq 0$, let C_R denote the circle $\{z : |z| = R\}$ and let μ_R denote the normalized Lebesgue measure of C_R. For $f \in L_\infty(\mathbb{R})$, let Sp(f) denote the spectrum of f. Finally, let us denote by $L_p^{(r)}(\mathbb{R}^2)$, $1 \leq p \leq \infty$, the radial functions in $L_p(\mathbb{R}^2)$.

Problems (i) and (ii) are solved by the following:

<u>Theorem 1</u>. Every right invariant w*-closed, non trivial subspace of $L_\infty(M(2))$ contains a function $e^{im\alpha}g(z) \neq 0$, $m \in Z$. The right invariant, w*-closed subspace generated by g(z), is minimal, if and only if, $Sp(g) \subseteq C_R$ for some $R \geq 0$.

<u>Proof</u>: Let f be a continuous function in the right invariant,

w*-closed, non-trivial subspace M. Then for a suitable $m \in Z$, the function

$$\int_0^{2\pi} f(e^{i(\alpha+\theta)},z)e^{im\theta}d\theta = e^{im\alpha}\int_0^{2\pi} f(e^{i\theta},z)e^{-im\theta}d\theta = e^{im\alpha}g(z)$$

is non-zero and belongs to M.

Moreover, if $g(z) \neq C$, then for a suitable $h \in L_1(\mathbb{R}^2)$ of the form $h(w) = h_1(r)e^{im\theta}$ $(w = re^{i\theta})$ the function

$$e^{im\alpha}\int_{\mathbb{R}^2} g(z - e^{i\alpha}w)h(w)\,dw = \int_{\mathbb{R}^2} g(z-w)h_1(r)e^{-i\theta m}\,dw = g_1(z)$$

is non-zero and belongs to M.

To prove the second part of the theorem, let $\lambda_1,\lambda_2 \in Sp(f)$ where $|\lambda_1|<|\lambda_2|$. For a radial function $h \in L_1(\mathbb{R}^2)$ whose spectrum lies in the closed annulus $\{w : r_1 \le |w| \le R$ where $|\lambda_1| < r_1 < |\lambda_2| < R$, we have

$$\int_{\mathbb{R}^2} g(z-e^{i\alpha}w)h(w)\,dw = \int_{R^2} g(z - w)h(|w|)\,dw = g^*(z) \neq 0.$$

The function $g^*(z)$ belongs to the w*-closed subspace V spanned by the right translates of g. However, for $K(z) \in L_1(\mathbb{R}^2)$ such that $\operatorname{Supp} \hat{K} = \{w : |\lambda_1| \le |w| \le r_3\}$ where $|\lambda_1| \le r_3 < r_1$, and satisfies $\int g(z)K(z)\,dz \neq 0$, we have

$$\int_{M(2)} g(z - e^{i\alpha}x)K(z)\,d\mu = 0 \quad \text{for every } x \in \mathbb{C}.$$

Hence $g \notin V$ and consequently, V is not minimal. Let $g(z) \in L_\infty(\mathbb{R}^2)$, $g \neq 0$ such that $Sp(g) \subseteq C_R$ for some $R > 0$. Suppose now that $p(z) \in W$, $p(z) \neq 0$, where W is the w*-closed subspace spanned by right translates of g. To derive the minimality of W, we will verify that $p(z) = Cg(z)$ for some $C \in \mathbb{C}$.

Let $\Phi_\tau(w)$, $\tau \varepsilon \Gamma$, be a net in $L_1(\mathbb{R}^2)$ such that

$$\int_{\mathbb{R}^2} g(z - e^{i\alpha}w)\Phi_\tau(w)\,dw \underset{M(2)}{\overset{w^*}{\rightarrow}} p(z).$$

Hence, we have

$$\frac{1}{2\pi}\int_0^{2\pi}\int_{\mathbb{R}^2} g(z - e^{i\alpha}w)\Phi_\tau(w)\,dw\,d\alpha \underset{M(2)}{\overset{w^*}{\rightarrow}} p(z)$$

and

$$(2) \qquad \int_{\mathbb{R}^2} g(z - w)\Phi^*_\tau(|w|)\,dw \underset{R^2}{\overset{w^*}{\rightarrow}} p(z)$$

where

$$(3) \qquad \Phi^*_\tau(|w|) = \frac{1}{2\pi}\int_0^{2\pi}\Phi_\tau(e^{-i\alpha}w)\,d\alpha.$$

But for every $\tau \varepsilon \Gamma$

$$\int_{\mathbb{R}^2} g(z - w)\ \Phi^*_\tau(|w|)\,dw = \hat{\Phi}^*_\tau(R)g(z).$$

Therefore $p(z) = Cg(z)$, which completes the proof.

Section 2

Our main result is the following:

Theorem 2. There exists a right invariant, w*-closed, non-zero subspace of $L_\infty(M(2))$, which does not contain any minimal, non-zero, right invariant, w*-closed subspace.

The proof of the theorem will be accomplished in several lemmas.

<u>Lemma 3</u>. Let f_1 and f_2 be in $L_\infty^{(r)}(\mathbb{R}^2) \cap L_1^{(r)}(\mathbb{R}^2)$ such that $\hat{f}_1$ is a constant d in a closed annulus R^*. Let C_R, $R>0$ be a circle contained in the interior of R^*. If Φ_τ, $\tau \in \Gamma$, is a net in $L_1^{(r)}(\mathbb{R}^2)$ such that

$$f_i * \Phi_\tau \underset{\tau}{\overset{w^*}{\to}} a_i \hat{\mu}_R \quad (i = 1,2),$$

then we have

$$(4) \qquad a_1 \hat{f}_2(R) = a_2 d$$

<u>Proof</u>: We may assume that $\operatorname{Supp} \hat{\Phi}_\tau \subseteq R^*$ for every $\tau \in \Gamma$. Hence

$$f_1 * \Phi_\tau = d\Phi_\tau \quad \text{for any } \tau \in \Gamma$$

Suppose that $d \neq 0$. Then $\Phi_\tau \overset{w^*}{\to} \frac{a_1}{d} \hat{\mu}_R$ and $f_2 * \Phi_\tau \overset{w^*}{\to} \frac{a_1}{d} \hat{f}_2(R) \hat{\mu}_R$

If $d = 0$, then $f * \Phi_\tau = 0$ for any $\tau \in \Gamma$ and we have $a_1 = 0$. This completes the proof of the Lemma.

<u>Lemma 4</u>. Let $T_H(x)$ be the function

$$T_H(x) = \begin{cases} H(x-1) & 1 \leq x < 2 \\ H & 2 < x \leq 3 \\ H(4-x) & 3 < x \leq 4 \\ 0 & \text{elsewhere} \end{cases}$$

If $f_\lambda(r) = T_H(r)e^{i\lambda r}$, $f_\lambda(r) \in L^{(r)}(\mathbb{R}^2)$, $\lambda \in \mathbb{R}$

then

$$\lim_{\lambda\to\infty} \|\hat{f}_\lambda\|_{L_\infty} = 0.$$

Proof: The lemma is an immediate consequence of the local isomorphism between the algebras of Fourier transforms of $L_1^{(r)}(\mathbb{R}^2)$ and $L_1(\mathbb{R}, 1+|x|^{1/2})$ [5].

The main tool used in proving Theorem 2 lies in the following lemma.

Lemma 5. Let $\{d_n\}_{n=0}$ be any sequence of positive numbers. Then there exists a sequence $\{g_n(r)\}_{n=0}^{\infty}$ in $L_\infty^{(r)}(\mathbb{R}^2)$ with the following properties:

(I) $||g_n||_{L_\infty} \leq d_n \quad (n = 1,2,...,)$

(II) $Sp(g_n) \subseteq \{z : 2 \leq |z| \leq 3) \quad (n = 0,1,...,)$

(III) If for some net $\Phi_\tau (\tau \in \Gamma)$ in $L_1^{(r)}(\mathbb{R}^2)$ and for some $R > 0$ we have

$$g_n * \Phi_\tau \underset{\tau}{\overset{w*}{\to}} c_n \hat{\mu}_R \quad \text{for } n = 0,1,...,$$

such that $\{c_n\}_{n=0}^{\infty} \in \ell_\infty$, then $c_n = 0$ for $n = 0,1,...,$.

Proof: Let $\psi \in L_1^{(r)}(\mathbb{R}^2)$, $||\psi||_{L_1} = 1$, such that $\operatorname{Supp} \hat{\psi} \subset \{z : 2<|z|<3\}$. Let g_n^* be the sequence defined by $g_0^*(r') = T_1(r')$ and $g_n^*(r') = T_n(r')e^{i\lambda_n r'}$ where the $\lambda_n \in \mathbb{R}$ satisfy $||g_n^*||_{L_\infty} \leq d_n$ $(n = 1,2,...,)$. Next, let g_n be defined by $g_n = g_n^* * \psi$ $(n = 0,1,...,)$. Obviously, g_n satisfies (I) and (II). To prove property (III), let R, $R > 0$, be such that $C_R \subset \operatorname{Supp} \hat{\psi}$, and let Φ_τ, $\tau \in \Gamma$, be a net in $L_1^{(r)}(\mathbb{R}^2)$ such that

$$(5) \qquad g_n * \Phi_\tau \underset{\tau}{\overset{w*}{\to}} c_n \hat{\mu}_R \quad c_n \hat{\mu}_R \qquad (n = 0,1,...,),$$

where $\{c_n\}_{n=o}^{\infty} \varepsilon \ell_\infty$.

From (5) we have $g_n^* * (\psi * \Phi_\tau \underset{\tau}{\overset{w*}{\to}} c_n \hat{\mu}_R$. By Lemma 3 we deduce that $c_n = c_o n e^{i\lambda_n R}$ $(n = 1,2,...,)$ which implies that $c_n = 0$ for each n.

The proof of Theorem 2. Suppose that every right invariant w*-closed, non zero subspace contains a minimal subspace.

From Theorem 1 it follows that for every $f(z) \varepsilon L_\infty M(2))$, $f(z) \neq 0$, there exist a net $\Phi_\tau \varepsilon L_1(\mathbb{R}^2)$, $\tau \varepsilon \Gamma$, such that

$$\int_{\mathbb{R}^2} f(z - e^{i\alpha} w)\Phi_\tau(w)dw \underset{M(2)}{\overset{w*}{\to}} g(z)$$

where $g \neq 0$ and $Sp(g) \subseteq C_R$, $R \geq 0$. Hence, we have

$$\int_{\mathbb{R}^2} f(z-w)\Phi_\tau^*(|w|)dw \underset{\mathbb{R}^2}{\overset{w*}{\to}} g(z)$$

where Φ_τ^* are defined in (3). In other words, for every $f \varepsilon L_\infty(\mathbb{R}^2)$, $f \neq 0$, there exist $R \geq 0$ and $g \varepsilon L_\infty(\mathbb{R}^2)$, $g \neq 0$, where $Sp(g) \subseteq C_R$, such that g is contained in the w*-closed subspace spanned by $\{f * \mu_R : R \geq 0\}$. To complete the proof, we will construct a non-zero function $f \varepsilon L_\infty(\mathbb{R}^2)$ such that the only function in the w*-closed subspace spanned by $\{f * \mu_R : R \geq 0\}$, whose spectrum lies in C_R, for some $R \geq 0$, is the zero function.

Let $b_n = ||K_n||_{L_1}$ where $K_n \varepsilon L_1(\mathbb{R}^2)$ such that $\hat{K}_n(r',\theta') = e^{i\theta' n}$ for $2 \leq r' \leq 3$ $(n = 0,1,...,)$.

Let $\{g_n\}_{n=o}^{\infty}$ be the sequence constructed in Lemma 5 where

$$d_n = \frac{|J_n(1)|}{2^n b_n}.$$

Here $J_n(x)$ denotes the nth Bessel function of the first kind.

Let $f \varepsilon L_\infty(\mathbb{R}^2)$ be given by

$$f(r,\theta) = \sum_{n=0}^{\infty} \frac{1}{J_n(1)} (g_n * K_n)(r,\theta) = \sum_{n=o}^{\infty} \frac{1}{J_n(1)} f_n(r)e^{in\theta}$$

where $f_n(r)$ is the nth Fourier coefficient of f.

Suppose that there exist a number R, $R>0$, and a function g, $g \neq 0$, $g \in L_\infty(\mathbb{R}^2)$ with $Sp(g) \subseteq C_R$, and a net Φ_τ, $\Phi_\tau \in L_1^{(r)}(\mathbb{R}^2)$, $\tau \in \Gamma$, such that

$$f * \Phi_\tau \overset{w^*}{\to} g.$$

From the explicit form of the nth Fourier coefficient of g as given in [6], we deduce that

$$\frac{1}{J_n(1)} [f_n(r)e^{in\theta}] * \Phi_\tau \underset{\tau}{\overset{w^*}{\to}} c_n \widehat{\mu_R e^{in\theta}} \quad (n = 0,1,...,),$$

where $\{J_n(1)c_n\}_{n=o}^{\infty} \in \ell_\infty$

Hence

$$(g_n * K_n) * \Phi_\tau \underset{\tau}{\overset{w^*}{\to}} c_n J_n(1)\, \mu_R e^{in\theta} \quad (n = 0,1,...,),$$

which yields

$$g_n * \Phi_\tau \underset{\tau}{\overset{w^*}{\to}} c_n J_n(1)\hat{\mu}_R$$

By Lemma 5 we deduce $c_n = 0$ $(n = 0,1,...,)$ and consequently $g = 0$.

REFERENCES

[1] G. Choquet, J. Deny, Sur l'équation de convolution $\mu = \mu * \sigma$, C.R. Acad. Sci. Paris, 250 (1960), 799-801.

[2] L. Elie and A. Rougi, Functions harmoniques sur certains groupes resolubles, C.R. Acad. Sci. Paris, Serie A (1975), 377-379.

[3] H. Furstenberg, A. Poisson formula for semi-simple Lie groups. Ann. of Math. 77 (1963), 335-386.

[4] H. Furstenberg, Translation - invariant cones of functions on semi-simple Lie groups, Bull. Amer. Math. Soc. 71 (1965), 271-326.

[5] M. Gatesoupe, Caractérisation locale de la sous algèbre fermée des functions radiales de $L^1(\mathbb{R}_n)$, Ann. Inst. Fourier, Grenoble 17 (1967) 93-197.

[6] C.S. Herz, Spectral synthesis for the circle, Ann. of Math. 68 (1958) 709-712.

[7] H. Leptin, Ideal theory in group algebras of locally compact groups. Inventiones Math. 31 (1976) 259-278.

[8] P. Mueller-Roemer, A tauberian group algebra, Proc. Amer. Math. Soc. 37 (1973) 163-166.

[9] Y. Weit, On Wiener's tauberian theorem for a non-commutative group, Ph.D. Dissertation, Hebrew University of Jerusalem, January 1977 (Hebrew).

University of Haifa, Haifa, Israel.